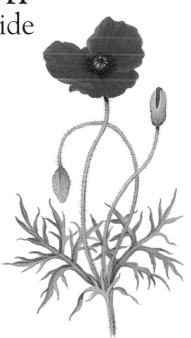

There are 47 individual Wildlife Trusts covering the whole of the UK and the Isle of Man and Alderney. Together The Wildlife Trusts are the largest UK voluntary organization dedicated to protecting wildlife and wild places everywhere – at land and sea. They are supported by 791,000 members, 150,000 of whom belong to their junior branch, Wildlife Watch. Every year The Wildlife Trusts work with thousands of schools, and their nature reserves and visitor centres receive millions of visitors.

The Wildlife Trusts work in partnership with hundreds of landowners and businesses in the UK. Building on their existing network of 2,250 nature reserves, The Wildlife Trusts' recovery plan for the UK's wildlife and fragmented habitats, known as A Living Landscape, is being achieved through restoring, recreating and reconnecting large areas of wildlife habitat.

The Wildlife Trusts also have a vision for the UK's seas and sea life – Living Seas, in which wildlife thrives from the depths of the oceans to the coastal shallows. In Living Seas, wildlife and habitats are recovering, the natural environment is adapting well to a changing climate, and people are inspired by marine wildlife and value the sea for the many ways in which it supports our quality of life. As well as protecting wildlife, these projects help to safeguard the ecosystems we depend on for services like clean air and water.

All 47 Wildlife Trusts are members of the Royal Society of Wildlife Trusts (Registered charity number 207238). To find your local Wildlife Trust visit wildlifetrusts.org

BLOOMSBURY

Concise
Wild Flower
Guide

B L O O M S B U R Y
LONDON · NEW DELHI · NEW YORK · SYDNEY

First published in 2010 by New Holland Publishers (UK) Ltd
This edition published in 2014 by Bloomsbury Publishing Plc

Bloomsbury Publishing Plc, 50 Bedford Square, London WC1B 3DP

www.bloomsbury.com

Bloomsbury Publishing, London, New Delhi, New York and Sydney

A CIP catalogue record for this book is available from the British Library
Library of Congress Cataloging-in-Publication Data has been applied for

Design by Alan Marshall

ISBN (print) 978-1-4729-0998-5

MIX
Paper from
responsible sources
FSC
www.fsc.org   FSC® C020056

Printed and bound in China by Leo Paper Group

1 0 9 8 7 6 5

Contents

Introduction

Meadows ablaze with wild flowers in bloom are the romantic ideal for many lovers of the countryside. When agriculture was less intensive, arable weeds thrived, roadsides had a more varied flora and 'unimproved' pasture had a wonderful variety of flowering plants. There is no doubt that there have been huge losses of habitat for wildlife, but prospects for wild flowers have recently improved. Conservationists began to draw the attention of land-users to the importance of biodiversity, and to their credit, local authorities and farmers have begun to take action. The challenge is to continue the restoration of the countryside beyond the confines of nature reserves.

Life Cycles of Flowering Plants

The flower is just one part of the life of a flowering plant. The cycle begins with the seed, is followed by the seedling, and grows to the point where it flowers and then fruits to release the seeds that start the cycle again. There is a huge variety of seed sizes. Some seeds can survive for many years, lying dormant in the soil, while others are so short-lived that they must germinate extremely quickly. Each plant may be able to produce a huge number of seeds, but few will germinate and even fewer will survive to flower. At every stage of its life a plant is vulnerable to predation, weather conditions and accidental destruction.

The roots of seedlings begin to develop before the leaves, and flowering plants are classified into two major orders depending on how the leaves of the seedlings develop. Monocotyledons have one seed leaf, or cotyledon, while dicotyledons have two.

Once the cotyledons have performed their function of protecting and nourishing the developing seedling, they wither to be replaced by true leaves. Secondary roots develop from the young plant's main roots to create an intricate network probing the soil for moisture and nutrients. Energy from sunlight is synthesized by chlorophyll, the green pigment in leaves, into carbohydrates, cellulose and starch. Growth throughout the plant requires the starches created in the leaves.

It is only when the plant is large enough that it will flower. Typically the flower consists of sepals, which form a calyx, within which are the petals. Sometimes both sepals and petals are similar, and together are called the perianth. Within the petals and sepals are the pollen-producing stamens, the male parts of the flowers. The female parts are the ovaries, each with one or more pollen-receptive stigmas often borne on a stalk-like part called a style. Seeds develop in the fertilized ovary, which ripens and turns into fruit. This can be juicy and berry-like, or dry.

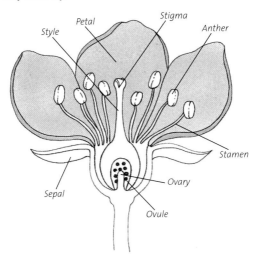

Plants are described according to their flowering and fruiting habits. The shortest-living are the annuals, which complete their lives within a single year and then die. Biennials germinate in their first year, survive through the winter, and in their second year flower, seed and die. Perennials flower from year to year, surviving the winter either above or below the ground.

Leaf Types

Typically the structure of a leaf is a stalk with a wide blade on either side of it. The blade is usually flat and may be simple or compound, consisting of several separate leaflets. Opposite leaves grow from the stem opposite each other, while alternate leaves grow from the opposite side of the stem, but not opposite each other. More than two leaves growing from the same point on the stem are described as whorled leaves. Simple leaves are found in a large number of shapes, including narrow-elliptical, broad-elliptical, narrow-lanceolate, kidney-shaped, arrow-shaped and ovate. Compound leaves consist of two or more separate leaflets. Pinnate leaves grow opposite each other, while trifoliate leaves grow in groups of three.

Flower Forms and Types

Regular flowers have similar petals and, even when there is an odd number, they give a regular appearance. Irregular flowers are those that are symmetrical only in that each side is similar. Some stems bear single flowers; these are known as solitary. Some flowers appear on spikes of several unstalked flowers on the stem. Clearly stalked individual flowers (for example growing on a spike) are called racemes, an example of which is the Lady's Smock. When each branch of the flower ends in a flower it is a cyme such as the Water Forget-me-not. Umbels are either flat-topped or domed flowerheads arising from the stem at the same place like an umbrella: this is typical of members of the carrot family such as Cow Parsley and Hemlock.

Adapting To Their Environment

Where a plant grows is the first clue to its identity – some species can thrive in a variety of habitats, while others are restricted. There are several factors that create habitats for plants: rainfall, temperature, soil types and proximity to the sea.

Coasts are often exposed to wind and salt spray, but there may be very little fresh water, so the plants that survive there must have the ability to retain moisture in their leaves and cope with salt. Aquatic plants grow in water. The flowers of the Common Water-crowfoot float on the surface, while those of the Arrowhead protrude above the surface of the water. In the acidic peat bogs the plant community includes the Broad-leaved Sundew, an insectivore that traps and absorbs small insects.

Open habitats cover heathland and grassland. Heathlands' main characteristics are acidic soils where woodlands were cleared many years ago by humans. The upland heaths are known as moorland. Grassland includes arable farmland, acid grassland and calcareous grassland. Due to intensive agriculture arable land is dominated by cereal crops, but some arable weeds can survive, especially with the aid of conservation-friendly farming. Acid grassland has a different flora from grasslands on chalk and limestone, and ancient meadows that have not been 'improved' have a wealth of flowers. The hedges that border meadows also have their own plant communities. Broadleaved woodlands contain mature trees and distinct layers of shrub and ground flora.

The 180 species in this book are a very small selection of the northern European list of almost 2,000. They are grouped into families with shared characteristics. Those chosen are widespread, common and particularly attractive – and likely to be easily spotted on walks, for which this portable little book is perfect.

Mistletoe
Viscum album

SIZE AND DESCRIPTION Up to 1m in diameter. A woody evergreen perennial parasite found on the branches of mature, usually deciduous trees, especially apples, limes and poplars. Forms spherical clumps. Branches are evenly forked and 30–100cm in length. Leaves are strap-shaped, leathery and yellowish-green, and borne in opposite pairs. Flowers are tiny (2–3mm in diameter) and green with four minute petals; male and female flowers grow on separate plants. Berries are white and sticky, and are dispersed by birds.

FLOWERING TIME November–February.

DISTRIBUTION Widespread throughout central and southern Europe, avoiding upland areas.

HABITAT Associated with open woodland in which the host tree species flourish.

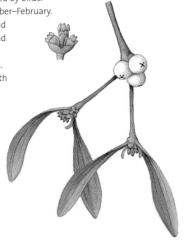

Hop
Humulus lupulus

SIZE AND DESCRIPTION Up to 6m tall. A twining climbing plant that uses other species for support. Stems are covered in stiff hairs. Rough-textured palmate leaves are divided into 3–5 leaflets. Male and female flowers are borne on separate plants. Male flowers are carried in branched clusters; female flowers are pendant cone-like hops, green at first but ripening to brown.

FLOWERING TIME August–September.

DISTRIBUTION Widespread in lowland Europe.

HABITAT Common in hedgerows and scrub habitats; also widely cultivated.

Common Nettle
Urtica dioica

SIZE AND DESCRIPTION Up to 1.5m tall.
A coarse upright perennial that is
covered with stinging hairs and has
tough yellowish roots. Stems bear
pairs of opposite leaves that are
ovate and toothed. Flowers are
small with four greenish petals;
borne in pendant axillary spikes;
male and female flowers grow on
separate plants. Female flowers
with paintbrush-like stigma.
FLOWERING TIME June–August.
DISTRIBUTION Widespread in
lowland Europe.
HABITAT Hedgerows,
woodlands and
disturbed ground
near habitation.
Thrives best on
nitrogen-
enriched soils.
Forms extensive
patches in
suitable locations.

Redshank
Persicaria maculosa

SIZE AND DESCRIPTION Up to 80cm tall. An upright or sprawling annual with red-tinged and much-branched stems. Leaves are narrow, oval and hairless, sometimes tinged red and invariably showing a dark central spot or smudge. Basal sheath of leaf has a hairy margin. Pinkish-red flowers are borne in dense axillary and terminal spikes.
FLOWERING TIME May–October.
DISTRIBUTION Widespread in lowland central and western Europe.
HABITAT Damp soils on cultivated and disturbed ground. Often found at the marshy margins of shallow lakes.

Common Sorrel
Rumex acetosa

SIZE AND DESCRIPTION Up to 60cm tall. An upright perennial. Leaves are dark green and shaped like an arrowhead; the lower ones are stalked while the upper ones are stalkless and clasp the stem. Leaves taste of vinegar. Flowers are reddish and borne in upright slender spikes. Occasionally, the whole plant may be tinged red.

FLOWERING TIME May–July.

DISTRIBUTION Widespread and common in much of Europe except far south.

HABITAT Wide range of grassy habitats, from meadows and woodland rides to coastal cliffs and dunes.

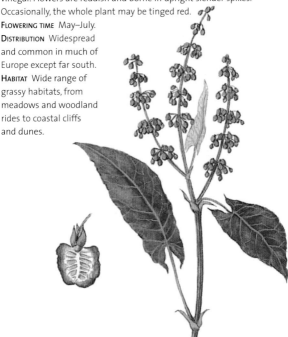

Water Dock
Rumex hydrolapathum

Size and description Up to 2m tall. A much-branched perennial. Leaves are 80cm or more in length, tough and narrow-oval in outline. Pale green flowers are borne on tall dense spikes.

Flowering time July–September.

Distribution Locally common in suitable habitats in lowland Europe.

Habitat Wetland areas, typically growing beside rivers and lakes, and sometimes rooted in shallow muddy water margins.

Broad-leaved Dock
Rumex obtusifolius

SIZE AND DESCRIPTION Up to 1m tall. A robust upright perennial. Stalked leaves are broadly oval, heart-shaped at their bases and 20cm or more In length; veins on undersides are hairy. Flowers are small and greenish, and borne in whorls along upright flower spikes, which are leafy at the bases.

FLOWERING TIME June–October.

DISTRIBUTION Widespread and often abundant in suitable habitats in lowland Europe.

HABITAT Disturbed ground such as field margins, tracks and waste ground.

Good-King-Henry
Chenopodium bonus-henricus

SIZE AND DESCRIPTION Up to 50cm tall. An upright perennial. Typically greenish but sometimes tinged red. Leaves are up to 10cm long and triangular, the lower ones most noticeably so; the leaf surface is mealy when young but becomes smooth and green with age. Stems sometimes show red lines. Flowers are small, reddish and borne on leafless spikes.

FLOWERING TIME May–August.

DISTRIBUTION Widespread in lowland Europe.

HABITAT Waste places and cultivated ground; often common on arable land.

Fat-hen
Chenopodium album

SIZE AND DESCRIPTION Up to 1.5mm tall. A striking upright annual. Dark green colour of plant is usually masked by covering of powdery white meal. Stems sometimes have reddish streaks. Leaves narrow-oval to diamond-shaped, and 20–70cm long; leaf margins are often toothed but seldom lobed. Flowers are whitish and borne on open leafy spikes.

FLOWERING TIME June–October.

DISTRIBUTION Widespread in lowland Europe and often abundant in suitable locations.

HABITAT Waste ground and disturbed arable land.

Three-nerved Sandwort
Moehringia trinervia

SIZE AND DESCRIPTION Up to
5cm tall. A rather delicate
downy annual with a
straggling or trailing
habit. Easily overlooked
among larger plants.
Leaves are oval and pointed
with three (sometimes five)
distinct veins. Flowers are
5–6mm across with five
white petals that are shorter
than the five green sepals.
Flowers are borne on slender stalks.

FLOWERING TIME May–June.

DISTRIBUTION Widespread throughout Europe, but
only locally common in suitable habitats.

HABITAT Rich damp soils in relatively undisturbed woodland.

Common Chickweed
Stellaria media

SIZE AND DESCRIPTION Up to 90cm tall. A straggly and much-branched annual. Leaves are oval, fresh green and borne in opposite pairs; upper leaves are unstalked. Flowers are 5–10mm across with five deeply divided white petals that are shorter than the five green sepals. Flowers are borne on stalks.

FLOWERING TIME Mainly July–November, but can be in flower in any month.

DISTRIBUTION Widespread in Europe; often considered a persistent weed.

HABITAT Disturbed ground and cultivated soils.

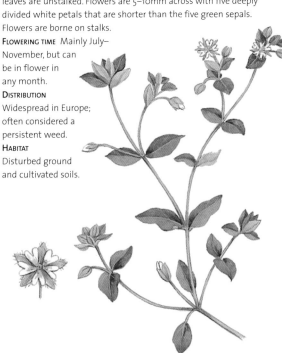

Lesser Stitchwort
Stellaria graminea

SIZE AND DESCRIPTION Up to 50cm tall.
A straggly perennial often found
growing through grasses. Leaves are
long, narrow and fresh green, closely
resembling those of grasses.
Flowers are 5–15mm across with
five deeply cleft white petals and five
green sepals of a similar length.
FLOWERING TIME May–August.
DISTRIBUTION Widespread throughout
lowland Europe except far south.
HABITAT Grassy places such as meadows
and woodland rides with acid soils.

Greater Stitchwort
Stellaria holostea

SIZE AND DESCRIPTION Up to 60cm tall. A perennial with four-angled stems that turn upwards from a slender weak base. Leaves are long, narrow and fresh green, paired on the stems and stalkless, with a distinctive rough edge.
Flowers are 20–30mm across and white with five distinctive petals, each of which is split to about halfway down.

FLOWERING TIME April–June.

DISTRIBUTION Widespread in Europe in suitable locations.

HABITAT Woods and hedges.

Sticky Mouse-ear
Cerastium glomeratum

SIZE AND DESCRIPTION Up to 40cm tall. A stickily hairy annual. Leaves are oval, pointed and borne in opposite pairs. Stems are often tinged red. Flowers are 10–15mm across, seldom open fully, and have five deeply notched white petals and five extremely hairy green sepals. Flowers are borne in tight terminal clusters.

FLOWERING TIME April–October.

DISTRIBUTION Widespread in lowland Europe except for north-east; often abundant in suitable locations.

HABITAT Disturbed habitats including fields, lawns and sides of roads.

White Campion
Silene latifolia

SIZE AND DESCRIPTION Up to 1m tall. A much-branched upright perennial that is stickily hairy all over. Leaves are oval, usually stalkless and borne in opposite pairs. Flowers are 25–30mm across with five white petals that are deeply notched. Flowers are borne in loose clusters.

FLOWERING TIME May–October.

DISTRIBUTION Widespread and locally common in lowland Europe.

HABITAT Hedgerows and grassy verges; also recently disturbed and cultivated ground.

Ragged-Robin
Lychnis flos-cuculi

SIZE AND DESCRIPTION
Up to 80cm tall. A distinctive
upright perennial. Stems are
rough and sometimes
branched. Stem
leaves are narrow,
rough, grass-like, stalkless
and borne in opposite pairs.
Basal leaves are stalked and
oblong. Flowers are pinkish-
red and ragged-looking with
five petals, each divided into
four lobes.

FLOWERING TIME May–August.

DISTRIBUTION Widespread and locally common in Europe; least common
in south. Has declined due to land drainage.

HABITAT Damp habitats including fens, water meadows and damp
woodland rides.

Red Campion
Silene dioica

SIZE AND DESCRIPTION
Up to 1m tall. A
downy upright
biennial or
perennial. Leaves
are oval, hairy and
borne in opposite pairs;
those on stems are usually
stalkless. Flowers are
20–25mm across, usually
deep pink but sometimes
much paler, with five petals. Male and
female flowers are borne on separate plants.
FLOWERING TIME Mainly March– November, but
can flower in any month.
DISTRIBUTION Widespread in lowland Europe; least common in warmer
southern areas.
HABITAT Wide variety of grassy habitats including woodland rides,
meadows and roadside verges.

Greater Sea-spurrey
Spergularia media

SIZE AND DESCRIPTION Up to 10cm tall. A robust fleshy perennial. Leaves are fleshy and bristle-tipped, paired on stem and unstemmed. Flowers are up to 60mm across, pink or whitish, growing on loose and widely branched heads.

FLOWERING TIME June–September

DISTRIBUTION Widespread throughout Europe in suitable locations.

HABITAT Muddy or sandy parts of salt marshes.

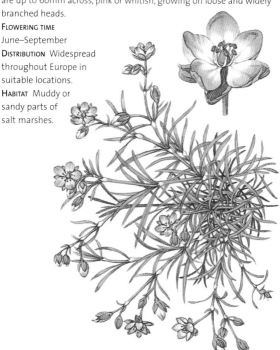

Marsh-marigold
Caltha palustris

SIZE AND DESCRIPTION Up to 60cm tall.
A striking hairless perennial with a creeping
habit. Basal leaves are kidney-shaped, dark
green and borne on long stalks. Stem leaves
are smaller and more rounded, with shorter
stalks. Flowers are 20–30mm across and
comprise five yellow sepals but no petals;
borne in loose clusters.

FLOWERING TIME March–July.

DISTRIBUTION Widespread in most of Europe but
essentially absent from Mediterranean region.

HABITAT Marshes, wet woodlands and fens.

Wood Anemone
Anemone nemorosa

SIZE AND DESCRIPTION Up to 30cm tall.
A low-growing hairless perennial. Stem
leaves are long-stalked and divided into
three lobes, each further divided. Flowers
are up to 40mm across with 5–10 whitish
or pinkish petal-like sepals. They are
solitary and borne on upright stalks.
FLOWERING TIME March–May.
DISTRIBUTION Widespread in much of
Europe, but absent from most of
Mediterranean.
HABITAT Open woodlands, often forming
extensive carpets; also alpine meadows.

Meadow Buttercup
Ranunculus acris

SIZE AND DESCRIPTION Up to 1m tall. A distinctive and familiar downy perennial. Leaves are divided into 3–7 segments, each of which is oval or wedge shaped, toothed and further divided.

FLOWERING TIME April–October.

DISTRIBUTION Widespread across most of Europe.

HABITAT Grassy areas like meadows and roadside verges. Often abundant in favourable locations.

Pasque Flower
Pulsatilla vulgaris

SIZE AND DESCRIPTION Up to 12cm tall.
An attractive low-growing hairy
perennial. Basal leaves are
especially hairy when they first
appear; they are much-divided into
feathery segments that do not
expand until the flowers
open. Bell-shaped flowers
are 55–85mm across
and have six
purple segments
that surround
numerous yellow stamens.
FLOWERING TIME March–May.
DISTRIBUTION Widespread but
local in central and north-west
Europe. Range much reduced
by changes in agriculture.
HABITAT Undisturbed ancient
grassland situated on
lime-rich soils.

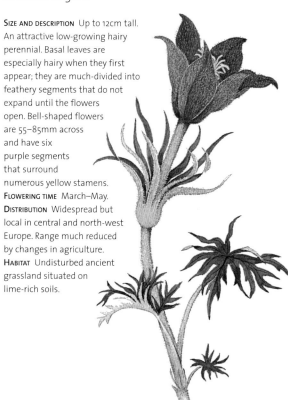

Traveller's-joy
Clematis vitalba

SIZE AND DESCRIPTION Up to 30m long. A vigorous woody climber with a rambling habit. Often smothers shrubs through which it grows by end of autumn. Leaves are pinnate and divided into pointed leaflets. Flowers are 20mm across and greenish-white; borne in loose clusters. Fruits are reddish, feathery and distinctive.

FLOWERING TIME July–September.

DISTRIBUTION Widespread in much of central, southern and western Europe.

HABITAT Hedgerows, woodland margins and scrub habitats; grows almost exclusively on lime-rich soils.

Common Water-crowfoot
Ranunculus aquatilis

SIZE AND DESCRIPTION Up to 1m long. An attractive annual or perennial; by end of summer it may blanket the surface of the wetland habitats where it grows. Surface leaves are rounded and toothed, while submerged ones are finely divided and thread-like. Flowers are 12–20mm across with five white petals.
FLOWERING TIME April–August.
DISTRIBUTION Widespread throughout most of lowland Europe.
HABITAT Slow-flowing streams and rivers, and still waters.

Common Meadow-rue
Thalictrum flavum

SIZE AND DESCRIPTION Up to 1m tall.
A striking upright perennial. Fern-like
leaves are divided into lobed leaflets.
Flowers comprise four small whitish
sepals and numerous protruding and
erect yellow stamens, the latter
resulting in a feathery appearance.
Flowers are borne in dense clusters.
FLOWERING TIME June–July.
DISTRIBUTION Widespread in most of
Europe, but usually distinctly local.
HABITAT Damp lowland habitats,
typically on base-
rich soils; also
locally in
mountains.

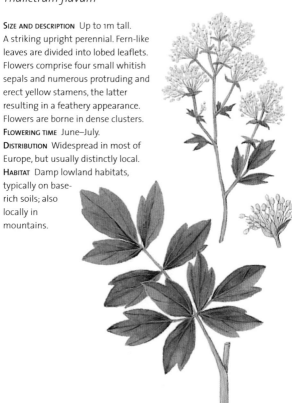

Lesser Celandine
Ranunculus ficaria

SIZE AND DESCRIPTION Up to 30cm tall. A low-growing perennial that can form extensive carpets. Glossy dark green leaves are heart- or kidney-shaped and borne on long stalks; may appear variegated. Flowers are 15–30mm across and borne on long stems; comprise three sepals and 8–12 yellow petals. They open fully only in bright sunshine.

FLOWERING TIME March–May.

DISTRIBUTION Widespread across most of Europe and sometimes locally abundant.

HABITAT Open woodland and hedgerows.

Yellow Horned-poppy
Glaucium flavum

SIZE AND DESCRIPTION Up to 90cm tall. A branching clump-forming biennial or perennial. Leaves are blue-green, the basal ones being deeply pinnately lobed and arranged in a rosette. Flowers are 60–80mm across and comprise two sepals and four papery bright yellow petals. The seed capsule is up to 30cm long, slender and curved.

FLOWERING TIME April–September.

DISTRIBUTION Most suitable coasts around Europe; generally only locally common.

HABITAT Stable stretches of coastal sand and shingle.

Common Poppy
Papaver rhoeas

SIZE AND DESCRIPTION Up to 70cm tall. A distinctive hairy annual. Leaves are divided into narrow-toothed segments; only the lower leaves are stalked. Flowers are 60–90mm across with two green sepals that drop when the flower opens, and papery bright red petals. Fruit is a flat-topped capsule, the seeds released through pores beneath the cap.

FLOWERING TIME April–August.

DISTRIBUTION Widespread throughout Europe.

HABITAT Disturbed soil. In the absence of herbicides, often colours whole fields red.

Greater Celandine
Chelidonium majus

SIZE AND DESCRIPTION Up to 2m tall. A brittle-stemmed perennial. Compound or pinnate leaves are borne on lightly haired stems that exude an orange sap when cut. Sparsely clustered flowers are 20–30mm across, bright yellow and stalked, with four oval petals. Fruit is pod like, 4cm long and splits to release many tiny black seeds.

FLOWERING TIME May–August.

DISTRIBUTION Almost everywhere in Europe.

HABITAT Hedgerows, woodland margins, walls and weedy areas on roadside verges.

Common Fumitory
Fumaria officinalis

SIZE AND DESCRIPTION Up to 10cm tall. A scrambling annual with nearly upright or clambering stems. Stalked grey-green leaves, divided into narrow lobes, are arranged spirally on stem. Elongated pink flowers are up to 10mm long, and have crimson tips and four unequal petals; both inner petals are hidden by larger outer petals. Flowers are borne in upright clusters. Fruits are nut-like, nearly globular and single seeded.

FLOWERING TIME June–August

DISTRIBUTION Widespread in Europe in suitable habitats.

HABITAT Weedy patches in arable fields and gardens, and waste ground.

Water-cress
Rorippa nasturtium-aquaticum

SIZE AND DESCRIPTION Up to 60cm long. Usually creeping perennial with hollow stems that float and grow upwards to flower. Leaves are dark green and pinnate with rounded leaflets, the terminal one of which is heart-shaped. Flowers are white, 4–6mm across, four-petalled and arranged in loose terminal clusters.

FLOWERING TIME May–October.

DISTRIBUTION Common and widespread across Europe except far north.

HABITAT Ditches, brooks and streams in shallow fresh running water.

Garlic Mustard
Alliaria petiolata

SIZE AND DESCRIPTION Up to 1.2m tall. A patch-forming upright biennial. Leaves are fresh green and heart-shaped with toothed margins; they smell strongly of garlic when crushed. Flowers are 4–6mm across with four white petals; flowers borne in clusters at stem tips. Fruits consist of slender pods.

FLOWERING TIME April–July.

DISTRIBUTION Widespread in most of Europe in suitable habitats.

HABITAT Hedgerows, woodland rides and scrub habitats on lime-rich soils.

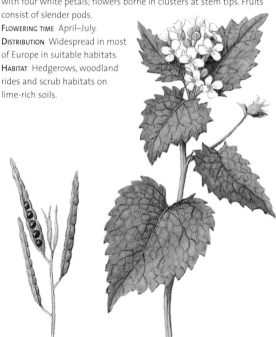

Shepherd's-purse
Capsella bursa-pastoris

SIZE AND DESCRIPTION Up to 40cm tall. A variable upright annual or biennial. Has a basal rosette of pinnately divided lobed leaves. Upper leaves are toothed and clasp the stem. Flowers are 2–3mm across with four white petals and four hairy green sepals; borne in terminal heads at stem tips. Fruits are heart-shaped, borne on erect stalks.

FLOWERING TIME Mainly April–October, but can flower in any month.
DISTRIBUTION Widespread throughout Europe.
HABITAT Disturbed ground in gardens, and on tracks and arable land.

Cuckoo Flower or Lady's Smock
Cardamine pratensis

SIZE AND DESCRIPTION Up to 55cm tall. An attractive upright perennial. Has a basal rosette of pinnately divided leaves with 1–7 pairs of rounded leaflets. Flowers are 12–20mm across with four whitish-pink or pale lilac petals; borne in open clusters at stem tips. Fruits consist of narrow upright pods up to 4cm long.

FLOWERING TIME April–July.

DISTRIBUTION Widespread in most of Europe in suitable habitats.

HABITAT Permanently damp ground in grassy habitats such as meadows, fens and woodland rides.

Charlock
Sinapis arvensis

SIZE AND DESCRIPTION
Up to 2m tall.
A striking upright
annual. Leaves are dark
green or purplish-green, and
coarsely toothed; lower leaves are
usually stalked, upper ones stalkless and
unlobed. Flowers are 15–20mm across
with four yellow petals; sepals are green
and down-turned; borne in dense terminal
clusters. Fruits consist of long pods.
FLOWERING TIME April–October.
DISTRIBUTION A native of southern Europe; introduced
elsewhere and now widespread across Europe.
HABITAT Arable land and disturbed waste ground.

Round-leaved Sundew
Drosera rotundifolia

SIZE AND DESCRIPTION Up to 8cm tall. A distinctive carnivorous plant. Basal rosette of rounded long-stalked leaves is covered with long sticky hairs that trap insects and envelop and digest them. Flowers are tiny, white and five-petalled, and open fully only in bright sunshine; borne terminally on upright curved-tipped stalks.

FLOWERING TIME June–August.

DISTRIBUTION Widespread but local due to very specific habitat requirements; absent from Mediterranean islands and southern Iberia.

HABITAT Waterlogged peaty soils in bogs and on heaths.

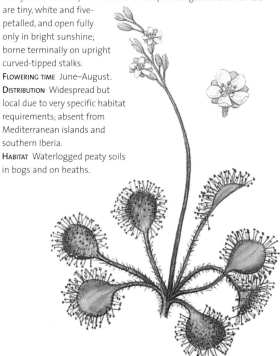

Biting Stonecrop
Sedum acre

SIZE AND DESCRIPTION Up to 10cm tall. A low-growing perennial that often forms mats. Leaves are succulent and crowded, spirally arranged and pressed close to stem. Star-shaped bright yellow flowers are 10–12mm across and have five pointed petals; borne on sparse and widely branched heads. Fruits are five pointed, and pods split to release numerous egg-shaped seeds.

FLOWERING TIME June–July.

DISTRIBUTION Widespread and locally common in Europe in suitable habitats.

HABITAT Warm and dry sunny cliffs, shingle, crevices in cobblestones, dunes, dry and grassy places, railway tracks and walls.

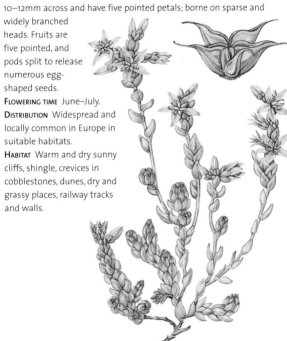

Meadow Saxifrage
Saxifraga granulata

SIZE AND DESCRIPTION Up to 50cm tall. An attractive hairy perennial. Leaves are basal, kidney-shaped and hairy with blunt teeth. Small brown bulbils are produced at leaf bases in autumn and give rise to new plants. Flowers are 20–30mm across with five white petals. Globular fruits split to release many seeds.

FLOWERING TIME May–July.

DISTRIBUTION Locally common in northern, central and western Europe. In southern and eastern Europe confined to cool upland regions only.

HABITAT Grassy meadows; also drier rocky sites, mostly on neutral to basic soils.

Meadowsweet
Filipendula ulmaria

SIZE AND DESCRIPTION Up to 2m tall.
A striking upright perennial. Leaves
are pinnately divided into pairs of
large toothed leaflets, interspersed
with pairs of smaller leaflets.
Flowers are 4–6mm across with
5–6 creamy white petals; borne
in long frothy sprays up to 25cm
in length. Small fruit is twisted
spirally and contains two seeds.
FLOWERING TIME June–September.
DISTRIBUTION Widespread
across most of Europe,
but essentially
absent from
Mediterranean.
HABITAT Damp soils
in meadows, marshes
and stream margins.

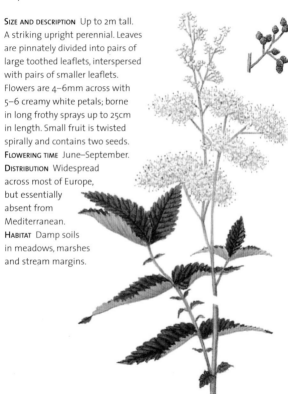

Agrimony
Agrimonia eupatoria

SIZE AND DESCRIPTION Up to 60cm tall. An upright perennial with slender hairy stems. Leaves are toothed and short-stalked, and positioned in 3–6 pairs on alternate sides of stem. Five-petalled golden-yellow flowers are borne on upright spires. Fruits have many tiny hooks that cling to animals' coats and are thus transported over wide areas.

FLOWERING TIME June–August.

DISTRIBUTION Widespread and common in Europe in suitable areas.

HABITAT Grassy places, hedgerows and roadsides, preferably situated in sunny locations.

Field-rose
Rosa arvensis

Size and description Up to 1m tall. A clump-forming shrub with trailing or scrambling stems that carry curved thorns. Leaves are divided into 5–7 oval leaflets. Flowers are 30–50mm across with five white petals; borne in open clusters of 2–6 blooms. Fruit is a red ovoid hip.

Flowering time July–August.

Distribution Widespread in western and central Europe; scarce or absent from most northerly and southerly extremes.

Habitat Woodland margins, hedgerows and scrub habitats.

Dog-rose
Rosa canina

SIZE AND DESCRIPTION Up to 5m tall. A scrambling deciduous shrub with erect or arched stems that are covered in hooked thorns. Leaves are pinnate and divided into 5–7 leaflets. Flowers are 30–50mm across with five pink or whitish petals; often solitary but sometimes in small groups. Fruit is a 10–20mm-long, ovoid bright red hip.

FLOWERING TIME June–July.

DISTRIBUTION Widespread in Europe except far north.

HABITAT Hedgerows, woodland margins and scrub habitats.

Wood Avens
Geum urbanum

SIZE AND DESCRIPTION Up to 60cm tall.
A delicate and elegant hairy
perennial. Basal leaves are
pinnate and comprise 1–5 leaflets;
stem leaves are 3–5-lobed. Flowers
are 8–15mm across with five spreading
yellow petals and five green sepals; flowers
are erect at first but soon droop. Fruits have
hooked hairs for clinging to animal fur.

FLOWERING TIME June–August.

DISTRIBUTION Locally common in most of Europe in suitable habitats.

HABITAT Shady woodland and hedgerows.

Tormentil
Potentilla erecta

SIZE AND DESCRIPTION Up to 30cm tall.
A charming perennial. Upright and
creeping stems typically weave through
other low-growing vegetation. Leaves are unstalked and trifoliate, but
appear five-lobed due to two leaflet-like stipules at base. Flowers are
7–11mm across with four easily dislodged bright yellow petals.
FLOWERING TIME May–September
DISTRIBUTION Widespread across most of Europe.
HABITAT Grassy habitats including meadows, heaths and moors.

Silverweed
Potentilla anserina

SIZE AND DESCRIPTION Up to 80cm long. A distinctive creeping perennial. Long-stalked leaves are silvery-green and comprise 5–7 toothed and lobed leaflets. Leaves form a persistent basal rosette. Flowers are 5–20mm across with five bright yellow petals; borne on sprawling radiating stems that root at the nodes.

FLOWERING TIME June–September.

DISTRIBUTION Widespread and common across most of Europe except far south.

HABITAT Open habitats including bare grassy places, roadside verges, waste ground and coastal dunes.

Wild Strawberry
Fragaria vesca

SIZE AND DESCRIPTION Up to 30cm tall. A perennial with upright stems, or low and arching stems that root to make new plants. Coarsely serrated stalked leaves comprise three leaflets that are hairy beneath; leaves grow from base or are scattered on stem. Five-petalled white flowers are 12–18mm across. Fruits consist of tiny strawberries.

FLOWERING TIME April–July.

DISTRIBUTION Found throughout Europe.

HABITAT Woods, hedgerows and short grassland.

Bramble
Rubus fruticosus

Size and description Up to 3m tall. A perennial with very prickly arching woody stems that root at the tips. Leaves alternate on stem, and are divided into five oval, pointed and toothed leaflets that are hairy beneath. Flowers are 20–30mm across, pink or white, and five-petalled; borne on cylindrical heads. Fruits consist of berry-like segments, ripening from red to black, each enclosing a seed.

Flowering time June–September.

Distribution Widespread and common in most of Europe.

Habitat Scrub, hedgerows and woods.

Common Gorse
Ulex europaeus

SIZE AND DESCRIPTION Up to 2m tall. A densely branched evergreen shrub. Twigs are greenish and almost leafless but bear numerous sharp spines, each up to 25mm long. Flowers are 15–25mm long, bright yellow and coconut-scented.

FLOWERING TIME April–May.

DISTRIBUTION Widespread native of western Europe that has been planted and naturalized elsewhere.

HABITAT Heaths, rough grassland and coastal cliffs on acid soils.

Broom
Cytisus scoparius

SIZE AND DESCRIPTION Up to 2m tall. A much-branched deciduous shrub with five-angled ridged green stems. Stems typically erect but coastal forms are prostrate. Leaves are small and usually trifoliate. Bright yellow flowers are 20mm long and borne on young shoots.

FLOWERING TIME April–June.

DISTRIBUTION Widespread in most of Europe.

HABITAT Open woodland, heaths and coastal cliffs on dry acid soils.

Tufted Vetch
Vicia cracca

SIZE AND DESCRIPTION
Up to 2m tall.
An elegant
scrambling
perennial that
climbs by means of
tendrils and may
smother the
vegetation that it grows
through by late summer.
Leaves are pinnately divided
into 6–15 leaflets and end in branched
tendrils. Flowers are bluish-purple; borne in
stalked spikes up to 10cm long. Fruits are hairless brown pods.
FLOWERING TIME June–August.
DISTRIBUTION Widespread in most of Europe.
HABITAT Grassy and scrub habitats including meadows, waste ground,
arable fields, riverbanks and edges of woodland.

Common Vetch
Vicia sativa

Size and description Up to 75cm tall. A delicate-looking downy annual that scrambles through other low-growing vegetation. Leaves comprise 3–8 pairs of oval leaflets and end in tendrils. Flowers are 20–30mm long and pinkish-purple; borne in groups of 1–2. Fruits are oblong pods that split to release globular seeds.

Flowering time April–September.

Distribution Widespread as native species in most of lowland Europe except far north. Cultivated for fodder and naturalized in many areas.

Habitat All sorts of grassy habitats.

Meadow Vetchling
Lathyrus pratensis

Size and description Up to 50cm tall. A perennial with long angular stems that climb through low-growing shrubs. Leaves comprise a pair of narrow leaflets with a twining tendril. Up to 12 greenish-veined yellow flowers with five petals each are borne in a long stalked cluster; flowers are attractive to bees, bumblebees and wasps. Pods are 25–35mm long, ripening to black.

Flowering time May–August.

Distribution Found throughout Europe.

Habitat Grassy areas, scrub and hedges.

Common Restharrow
Ononis repens

SIZE AND DESCRIPTION
Up to 70cm tall. A robust
perennial undershrub with
spreading or upright hairy
stems. Plant has a foetid
smell when rubbed. Stems
may bear soft spines. Leaves are
stickily hairy and usually trifoliate with
oval leaflets. Flowers are 10–15mm long,
pink and white, and borne in clusters.
Fruits are small pods.

FLOWERING TIME July–September.

DISTRIBUTION Locally common in western, south-
western and central Europe.

HABITAT Dry grassy places on calcareous soils.

Common Bird's-foot-trefoil
Lotus corniculatus

SIZE AND DESCRIPTION Up to 35cm
tall. A perennial that may be
hairy or hairless and can have
a creeping or ascending habit.
Leaves are grey-green and
downy, comprising five leaflets
that appear trifoliate. Flowers
are 10–16mm long and usually
yellow, tinged reddish; borne
on stalked heads of 2–7
flowers. Fruits are slender
pods that are splayed
like birds' toes.

FLOWERING TIME
April–September.

DISTRIBUTION
Widespread and
often common in
most of Europe in
suitable habitats.

HABITAT Grassy
habitats; tolerates a
wide range of soils.

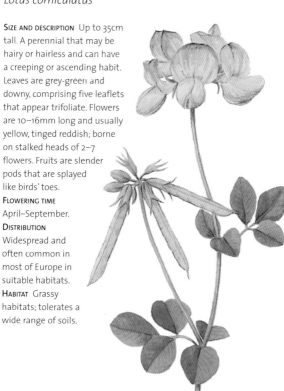

Black Medick
Medicago lupulina

SIZE AND DESCRIPTION Up to 20cm tall. A low-growing and often trailing downy annual; sometimes a short-lived perennial. Leaves are trifoliate with each leaflet having a point at the centre of its apex. Flowers are 2–3mm long, yellow and borne in packed clusters that are carried on long stalks. Pods are kidney-shaped and black when ripe.

FLOWERING TIME
April–October.

DISTRIBUTION Widespread in most of lowland Europe. Sometimes cultivated for fodder.

HABITAT Dry grassy places and disturbed ground.

Red Clover
Trifolium pratense

SIZE AND DESCRIPTION Up to 40cm tall. A perennial with almost upright stems. Leaves are trifoliate, the oval leaflets each bearing a crescent-shaped white mark. Flowers are pinkish-red, borne in dense, rounded and unstalked heads 20–30mm across at tip of stem. Each flower consists of a tube of five sepals and five petals. Flowers are attractive to bumblebees.

FLOWERING TIME May–September.

DISTRIBUTION Found throughout Europe.

HABITAT Grassy places including meadows, roadside verges and pastures.

White Clover
Trifolium repens

SIZE AND DESCRIPTION Up to 20cm tall.
A hairless perennial with prostrate
branching stems that root and form
extensive mats. Leaves are trifoliate and
borne on upright stems; leaflets are
ovate with a white mark and darker
veins. Flowers are borne in rounded heads
20mm across; flowers are usually whitish but
sometimes tinged with pink.

FLOWERING TIME
April–October.

DISTRIBUTION
Widespread across
most of Europe and
locally abundant.

HABITAT Grassy habitats.

Kidney Vetch
Anthyllis vulneraria

SIZE AND DESCRIPTION Up to 90cm tall. A silky and hairy annual, biennial or perennial. Leaves are pinnately divided into 1–7 pairs of narrow leaflets. Flowers are usually deep yellow, but may be purple, red or orange; borne in long-stalked, kidney-shaped clustered heads, 30mm across. Ripe pods are brown.

FLOWERING TIME April–September.

DISTRIBUTION Widespread throughout most of Europe.

HABITAT Grassy and rocky habitats on calcareous soils, from coastal cliffs to chalk downland and mountain pastures.

Wood-sorrel
Oxalis acetosella

SIZE AND DESCRIPTION Up to 10cm tall. A downy perennial with a creeping habit. Leaves are trefoil and shamrock-like; borne on long stalks in the form of a rosette. Leaves fold down at night. Flowers are solitary, 8–15mm across, borne on slender stalks. Petals usually white with purple veins; sometimes tinged purple. Fruit a five-ridged capsule that splits to release seeds.

FLOWERING TIME April–May.

DISTRIBUTION Widespread and locally common across Europe.

HABITAT Undisturbed woodland and hedgerows. Common under oaks and beeches.

Meadow Crane's-bill
Geranium pratense

SIZE AND DESCRIPTION Up to 80cm tall. A striking branched and hairy perennial. Leaves are divided into 5–7 ovate, deeply cut lobes. Flowers are 30–40mm across with five rounded bluish-violet petals with darker veins; borne in compact clusters that droop after flowering. Fruit a long-beaked capsule with each of five basal lobes breaking to release a seed, coiling back like a clock spring.

FLOWERING TIME
June–September.

DISTRIBUTION
Widespread and locally common in Europe, but rare in Mediterranean and northern regions.

HABITAT Sunny meadows and roadside verges, usually on base-rich soils.

Common Stork's-bill
Erodium cicutarium

SIZE AND DESCRIPTION Up to 30cm. Stickily hairy annual with stems that creep or are angled upwards. Leaves are mostly paired with lobed leaflets. Up to nine five-petalled pinkish-purple flowers form a long-stalked loose head. Fruits consist of a long-beaked capsule, each lobe containing a seed, detaching with a strip from top of beak to form a corkscrew-like coil, enabling the seed to anchor itself to the ground.

FLOWERING TIME
June–September.

DISTRIBUTION Across Europe.

HABITAT Dunes, waste ground, pathways and cultivated ground.

Herb-Robert
Geranium robertianum

SIZE AND DESCRIPTION Up to 30cm
tall. A straggling and rather
delicate annual. Plant is strong
smelling and may be tinged red.
Leaves are hairy and deeply cut into
3–5 lobes. Flowers are 12–15mm across,
five-petalled and pink; borne in loose
clusters on long stalks. Fruits are hairy
and end in long 'beak'.
FLOWERING TIME April–October.
DISTRIBUTION Widespread throughout
Europe except far north.
HABITAT Shady places such as rocky
banks, hedgerows and woodlands.

Fairy Flax
Linum catharticum

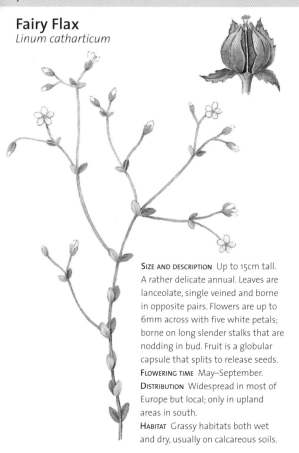

SIZE AND DESCRIPTION Up to 15cm tall. A rather delicate annual. Leaves are lanceolate, single veined and borne in opposite pairs. Flowers are up to 6mm across with five white petals; borne on long slender stalks that are nodding in bud. Fruit is a globular capsule that splits to release seeds.

FLOWERING TIME May–September.

DISTRIBUTION Widespread in most of Europe but local; only in upland areas in south.

HABITAT Grassy habitats both wet and dry, usually on calcareous soils.

Dog's Mercury
Mercurialis perennis

SIZE AND DESCRIPTION Up to 50cm
tall. An unbranched downy-hairy
perennial. Leaves are shiny, dark
green and ovate-lanceolate with
toothed margins; most are borne
on upper part of plant. Clusters of
rather insignificant greenish
flowers are borne on upright
spikes. Male and female flowers
are on separate plants.

FLOWERING TIME March–May.

DISTRIBUTION Widespread in most of
Europe except far north.

HABITAT Woodlands, especially under
oaks or beeches; may be abundant.

Sun Spurge
Euphorbia helioscopia

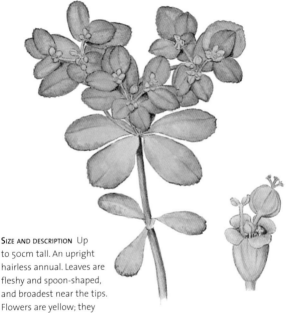

SIZE AND DESCRIPTION Up
to 50cm tall. An upright
hairless annual. Leaves are
fleshy and spoon-shaped,
and broadest near the tips.
Flowers are yellow; they
lack both petals and sepals, and instead comprise oval green glands.
Umbel has five green bracts. Fruits are smooth.

FLOWERING TIME April–November.

DISTRIBUTION Widespread across Europe, and often abundant in
suitable locations.

HABITAT Arable land, waste ground and cultivated soils.

Common Milkwort
Polygala vulgaris

SIZE AND DESCRIPTION Up
to 30cm tall. A delicate
perennial with a trailing
or upright habit. Leaves are
alternate; lower ones are oval, while
upper ones are narrow and pointed.
Flowers are 6–8mm long and can be blue, pink or white; borne in
clustered terminal spikes of 10–40 flowers.

FLOWERING TIME June–September.

DISTRIBUTION Widespread throughout most of Europe.

HABITAT Grassy habitats on all but the most acidic of soils.

Common Mallow
Malva sylvestris

Size and description
Up to 1.5m tall.
A perennial that may
be upright or spreading in
habit. Leaves are rounded at
the base but five-lobed on the
stem. Flowers are 25–40mm across
with five pink petals that have purple
veins and are hairy at the base.
Flowering time June–September.
Distribution Widespread and common
throughout Europe.
Habitat Grassy places such as meadows
and roadside verges; often thrives best
in disturbed soil.

Hairy St John's-wort
Hypericum hirsutum

SIZE AND DESCRIPTION Up to 1m tall.
An upright downy perennial
with round stems. Leaves are
hairy, oblong to elliptical, and
are borne in opposite pairs.
Flowers are 15mm across and
comprise five yellow petals; the
five sepals are shorter than the
petals and are pointed with
black glands on the margins.
FLOWERING TIME June–September.
DISTRIBUTION Widespread in most
of Europe except much of south
north-east.
HABITAT Damp grassy areas,
especially along woodland rides.

Common Dog-violet
Viola riviniana

SIZE AND DESCRIPTION Up to 15cm tall. An almost hairless perennial. Leaves are long-stalked, heart-shaped and blunt-tipped. Flowers are 10–13mm across. The blue-violet petals are broad and unequal, and marked with dark veins towards the flower centre; lower petal has a pale lilac spur 3–5mm long.

FLOWERING TIME March–May.

DISTRIBUTION Widespread in most of Europe except south-east.

HABITAT Woodland rides and grassy places.

Field Pansy
Viola arvensis

SIZE AND DESCRIPTION Up to 15cm tall. An annual with branched and nearly upright stems. Leaves are located on alternate sides of stem between paired, oblong or almost spoon-shaped leaf-like stipules with rounded teeth. Five-petalled flattish flowers are 10–15mm across and creamy white, sometimes with a yellow flush on lower petal.
FLOWERING TIME April–October.
DISTRIBUTION Widespread and common throughout Europe.
HABITAT Waste ground and cultivated ground.

Heartsease or Wild pansy
Viola tricolor

SIZE AND DESCRIPTION Up to 40cm
tall. A branching hairless or
downy-hairy annual, biennial or
perennial. Lower leaves are ovate,
while upper ones are oblong; the
stipules are deeply divided.
Flowers are 10–15mm across;
petals are unequal and yellow,
violet or bicoloured; lower one
bears a spur, 6mm long. Sepals
are longer than petals.
FLOWERING TIME April–November.
DISTRIBUTION Most of Europe.
HABITAT Cultivated ground
and grassland.

Purple-loosestrife
Lythrum salicaria

SIZE AND DESCRIPTION Up to
1.5m tall. A downy perennial.
Upright stems carry narrow
and unstalked leaves either as
opposite pairs or in whorls of
three. Flowers are 10–15mm
across and borne in tight
whorls creating a tall spike;
petals are reddish-purple and
there are 12 stamens. Fruit is
an egg-shaped capsule.
FLOWERING TIME June–August.
DISTRIBUTION Widespread across
Europe. Absent from far north.
HABITAT Damp ground, typically
beside water. Forms extensive
stands in suitable locations.

Rosebay Willowherb
Chamerion angustifolium

SIZE AND DESCRIPTION Up to 1.5m tall.
A showy upright perennial. Leaves
are lanceolate and arranged spirally
up the stem. Flowers are 20–30mm
across with four pinkish petals,
which are slightly unequal. Seeds
are small with a long plume of
silky white hairs, which assists
in wind dispersal.

FLOWERING TIME
June–August.

DISTRIBUTION Widespread
across most of Europe
except north.

HABITAT Wide range of
disturbed ground. Forms
large clumps.

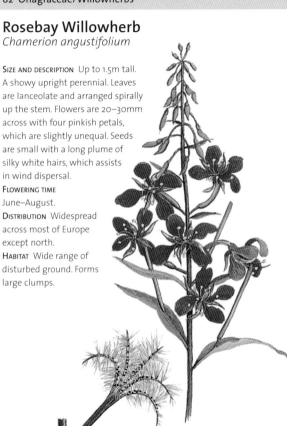

Enchanter's-nightshade
Circaea lutetiana

SIZE AND DESCRIPTION Up to 60cm tall.
A delicate creeping perennial.
Upright stems may be hairy, and
have opposite pairs of ovate pointed
leaves that are heart-shaped or
rounded at the bases. Flowers are
4–8mm across with two deeply
divided white petals; borne on a
loose spike above leaves. Hook-
shaped bristles cling to animals'
fur, assisting in seed dispersal.
FLOWERING TIME June-September.
DISTRIBUTION Widespread in most
of Europe except north-east.
HABITAT Woodlands and
shady gardens.

Great Willowherb
Epilobium hirsutum

SIZE AND DESCRIPTION Up to 2m
tall. A downy or hairy perennial.
Leaves are stalkless and
opposite, toothed, and
lanceolate to oblong. Flowers are 15–25mm
across with four pinkish-purple petals that are notched at the tips
and have pale centres. Fruit capsule splits to release plumed seeds.
FLOWERING TIME June–August.
DISTRIBUTION Widespread in most of Europe except far north.
HABITAT Damp soils in fens and marshes and on margins of rivers.
Often forms extensive and sizeable clumps.

Broad-leaved Willowherb
Epilobium montanum

SIZE AND DESCRIPTION Up to 80cm tall. Upright perennial that spreads scaly creeping stems and sends up tall flowering stems. Pointed, oval and toothed leaves are mostly paired, but sometimes in threes. Flowers are 6–10mm across, pale pink, with four narrowly oval, deeply notched equal petals. Capsule splits to release masses of plumed seeds.

FLOWERING TIME June–August.

DISTRIBUTION Widespread across Europe.

HABITAT Woods, hedges, among rocks and on cultivated ground.

Common Ivy
Hedera helix

SIZE AND DESCRIPTION Up to 20m tall. A self-clinging evergreen perennial climber with woody stems anchored by clinging roots; also carpets the ground. Short-stalked leaves are glossy dark green, with paler veins and 3–5 lobes. Five-petalled flowers are 5–8mm across, yellowish-green and borne in globular heads. Fruits are berries that darken to purplish-black.

FLOWERING TIME September–November.

DISTRIBUTION Widespread and common in Europe.

HABITAT Woods, hedgerows, scrub, rocks and walls.

Cow Parsley
Anthriscus sylvestris

SIZE AND DESCRIPTION Up to 1m tall. An upright downy perennial. Hollow ridged stem is purple and carries leaves that are 2–3 times pinnately divided. Flowers are small and white, and borne in flattened umbels that lack lower bracts. Fruits are dark brown.

FLOWERING TIME April–June.

DISTRIBUTION Most of Europe.

HABITAT Grassy places such as roadside verges, woodland margins and lanes. Forms extensive patches in suitable locations.

Sea-holly
Eryngium maritimum

SIZE AND DESCRIPTION Up to 60cm tall. An intriguing and distinctive perennial. Leaves are bluish-green, leathery and ovate with sharp spines that give them a holly-like appearance; basal leaves are stalked while stem leaves are stalkless. Flowers are small and blue; borne in globular umbels up to 20mm across, with spiny bracts below.

FLOWERING TIME June–September.

DISTRIBUTION Widespread in suitable areas on European coasts except in north.

HABITAT Stable and undisturbed coastal sand and shingle.

Hemlock
Conium maculatum

SIZE AND DESCRIPTION Up to 2m tall. A distinctive upright perennial, all parts of which are highly poisonous. Stem is hollow, ridged and blotched purple. Leaves are up to four times pinnately divided into fine leaflets. Flowers are small and white; borne in umbels up to 50mm across. Oval fruits have wavy or toothed ridges.

FLOWERING TIME June–August.

DISTRIBUTION Widespread in Europe except far north.

HABITAT Damp wayside ground, along river margins and on waste ground.

Hogweed
Heracleum sphondylium

SIZE AND DESCRIPTION
Up to 2m tall. An
extremely robust
perennial. Stem is stout, hollow and
hairy. Leaves are up to 60cm long, broad,
hairy and usually pinnately divided into
up to nine toothed segments. Flowers are small, off-white and with
unequal petals; borne in slightly domed umbels up to 200mm across.
Fruits are ridged with dark lines.

FLOWERING TIME April–November.

DISTRIBUTION Widespread except in Mediterranean and far north.

HABITAT Open grassy places like meadows and roadside verges.

Wild Carrot
Daucus carota

SIZE AND DESCRIPTION Up to 75cm tall. Upright or spreading hairy perennial that smells of fresh carrots and has solid rigid stems. Leaves are divided into slender lobed leaflets. Five-petalled flowers are white (pinkish in bud), 2–4mm across; borne in long-stalked umbels that are 70mm across, with central flower usually purple or red. Fruits are ridged and bear spines.

FLOWERING TIME June–August/September.

DISTRIBUTION Found almost everywhere in Europe.

HABITAT Meadows, wasteland, roadside verges and quarries; prefers dry and chalky soil.

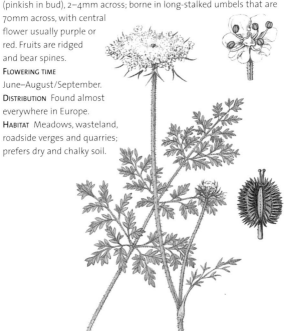

Ground-elder
Aegopodium podagraria

SIZE AND DESCRIPTION Up to 1m tall.
A creeping, patch-forming and hairless
perennial that spreads rapidly by
underground stems. Leaves are fresh
green and roughly triangular; lower
leaves have long three-angled stalks,
upper leaves have broad bases. Flowers
are 1–3mm across and white; borne
in compact umbrella-
shaped heads that are
up to 60mm across.
FLOWERING TIME May–July.
DISTRIBUTION Found almost
everywhere in Europe.
HABITAT Riverbanks, damp
woods, forest edges,
gardens and parks in
partial shade. Can be
invasive in gardens.

Scarlet Pimpernel
Anagallis arvensis

SIZE AND DESCRIPTION Up to 50cm long. A prostrate and almost hairless annual. Leaves are ovate to lanceolate; borne as opposite pairs along branching stems. Flowers are 12–15mm across with five lobes, and are carried singly on slender stalks; usually scarlet but sometimes pink or blue. Flowers close by early afternoon or at any time in dull weather. Fruits are globular capsules opening at top to release seeds.

FLOWERING TIME March–October.

DISTRIBUTION Widespread in most of Europe.

HABITAT Disturbed and cultivated ground. Can be locally abundant.

Cowslip
Primula veris

Size and description Up to 25cm tall.
A charming perennial. Leaves are hairy
and tapering, similar to those of
Primrose but more wrinkled; they form
a basal rosette. Flowers are orange-yellow,
bell-shaped and 8–15mm across; borne in drooping one-sided heads
on long, upright and naked stalks.

Flowering time April–June.

Distribution Widespread and locally common in much of Europe
except far north and south.

Habitat Grassland, open woodland rides and scrub on lime- and
chalk-rich soils.

Primrose
Primula vulgaris

Size and description Up to 20cm tall.
A clump-forming hairy perennial. Leaves are oval and tapering, up to
12cm long, and form a rosette. Delightful flowers are 20–30mm across
with five lobes that are usually pale yellow; flowers are solitary and
borne on long and hairy stalks arising from centre of leaf rosette.
Flowering time March–June.
Distribution Widespread and locally common in much of Europe.
Habitat Woodland rides and margins, shady meadows and hedgerows.

Bell Heather
Erica cinerea

SIZE AND DESCRIPTION Up to 50cm tall. A hairless evergreen undershrub. Leaves are narrow and needle-like; borne in whorls of three up wiry stems. Flowers are 5–6mm long, bell-shaped and purplish-red; appear in open clusters towards stem tips.

FLOWERING TIME June–September.

DISTRIBUTION Widespread in western Europe in suitable habitats, but also occurs more locally as far north as southern Scandinavia and as far south as northern Italy.

HABITAT Dry acid soils located on heaths and moors.

Common Heather or Ling
Calluna vulgaris

Size and description Up to 1.5m tall.
A much-branched dense undershrub.
Leaves are short, narrow and scale-like;
borne in four rows along stems. Flowers
are 4–5mm long, usually pinkish, but
sometimes almost white; borne on
long terminal spikes.
Flowering time July–September.
Distribution Widespread in much of
Europe in suitable habitats. Absent
from most of Mediterranean.
Habitat Acid soils on heaths and moors.
In suitable areas can form extensive
ground cover.

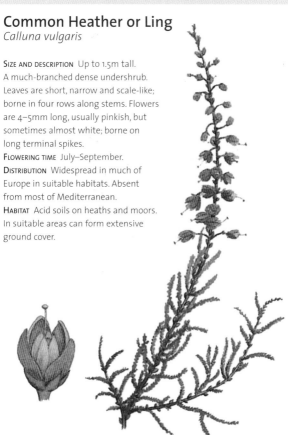

Bilberry
Vaccinium myrtillus

SIZE AND DESCRIPTION Up to 60cm tall. An upright deciduous shrub. Leaves are oval and bright green; borne on twigs that are green and three-angled. Flowers are 4–6mm across and greenish-pink, and eventually ripen to form an edible blue-black berry 6–10mm across.
FLOWERING TIME April–June.
DISTRIBUTION Widespread in much of Europe, but restricted to upland regions in south.
HABITAT Heaths, moors and woodlands on acid soils.

Thrift
Armeria maritima

SIZE AND DESCRIPTION Up to 60cm tall. A distinctive cushion-forming perennial. Leaves are narrow, grass-like and up to 15cm long, arising as numerous rosettes from woody base of plant. Flowers are pink and borne in heads 10–30mm across on slender, erect and leafless stalks.
FLOWERING TIME Mainly May–June, and occasionally April–October.
DISTRIBUTION Widespread and often locally abundant around coasts throughout Europe.
HABITAT Coastal cliffs and salt marshes; also occasionally grows on mountain-tops.

Common Centaury
Centaurium erythraea

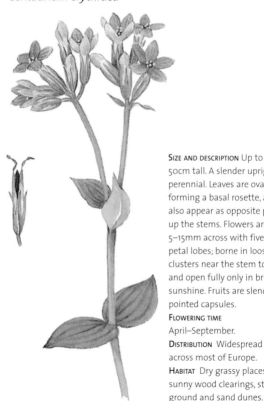

Size and description Up to 50cm tall. A slender upright perennial. Leaves are ovate, forming a basal rosette, and also appear as opposite pairs up the stems. Flowers are 5–15mm across with five pink petal lobes; borne in loose clusters near the stem tops and open fully only in bright sunshine. Fruits are slender pointed capsules.
Flowering time April–September.
Distribution Widespread across most of Europe.
Habitat Dry grassy places, sunny wood clearings, stony ground and sand dunes.

Cleavers
Galium aparine

SIZE AND DESCRIPTION Up to 1.5m tall.
A sprawling annual with backwards-
pointing bristles on the edges of the
stems to help secure their scrambling
progress through vegetation. Bristly and
stalkless spear-shaped leaves encircle
stems in rings of 6–8. Inconspicuous
greenish-white flowers are 2mm across
with four petals, and are grouped in
branched heads. Fruits have whitish
hooked bristles.
FLOWERING TIME June–August.
DISTRIBUTION Widespread and common
throughout Europe.
HABITAT Waste ground, hedge edges,
forests, arable fields, scrub, and towns
and villages.

Lady's Bedstraw
Galium verum

SIZE AND DESCRIPTION Up to 60cm
long. A trailing and sprawling
branched perennial. Leaves are
narrow and have rolled
margins; borne in whorls of
8–12 along four-angled stems.
Flowers are 2–3mm across with
four bright yellow petal lobes;
produced in dense clusters at
the ends of much-branched
stems. Two-lobed fruits
are dark brown.

FLOWERING TIME
June–September.

DISTRIBUTION Widespread
throughout much of Europe
except far north.

HABITAT Dry grassy habitats.

Hedge Bindweed
Calystegia sepium

SIZE AND DESCRIPTION Up to 3m tall. A vigorous perennial that climbs anti-clockwise around other plants; also spreads underground to make new plants. Leaves are arrow-shaped and up to 12cm long. Striking flowers are 30–40mm across, pure white and funnel-shaped, with five almost completely joined petals. Flowers attract moths.

FLOWERING TIME April–July.

DISTRIBUTION Found throughout Europe, though less common in north.

HABITAT Hedges and woods, and waste and cultivated ground.

Field Bindweed
Convolvulus arvensis

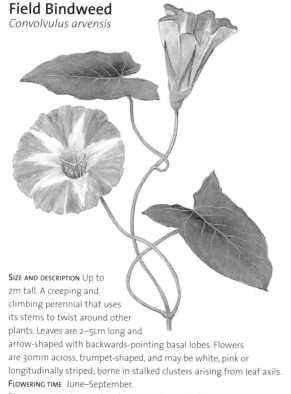

SIZE AND DESCRIPTION Up to 2m tall. A creeping and climbing perennial that uses its stems to twist around other plants. Leaves are 2–5cm long and arrow-shaped with backwards-pointing basal lobes. Flowers are 30mm across, trumpet-shaped, and may be white, pink or longitudinally striped; borne in stalked clusters arising from leaf axils.

FLOWERING TIME June–September.

DISTRIBUTION Widespread and common in most of Europe.

HABITAT Cultivated ground, roadside verges and disturbed land. Often considered an agricultural weed.

Bugloss
Anchusa arvensis

SIZE AND DESCRIPTION Up to 50cm tall. A roughly hairy annual with upright stems. Leaves are narrow with wavy margins, the lower ones being stalked, the upper ones clasping the stem. Flowers are 5–6mm across, have five bright blue petals and are produced in clusters.

FLOWERING TIME June–September.

DISTRIBUTION Much of Europe except parts of extreme north.

HABITAT Arable fields, heaths and vicinity of sea.

Viper's-bugloss
Echium vulgare

SIZE AND DESCRIPTION Up to 90cm
tall. A roughly hairy biennial. Leaves
are narrow and pointed, up to 15cm
long, and the basal ones are stalked.
Flowers are 15–20mm long, funnel-
shaped and bright blue, with pink
buds; borne in tall upright spikes
or shorter down-curved sprays.
FLOWERING TIME May–September.
DISTRIBUTION Widespread and
locally common throughout
most of Europe.
HABITAT Dry grassy habitats,
especially on chalky or sandy
ground, and often near coast.

Field Forget-me-not
Myosotis arvensis

SIZE AND DESCRIPTION Up to 25cm long. A downy branching annual with upright branched stems. Leaves are oblong or spear-shaped, with basal ones forming a rosette. Flowers initially have pink buds, which open as yellow-eyed blue flowers 5mm across. Dark brown fruits.

FLOWERING TIME April–September.

DISTRIBUTION Widespread and common except in north.

HABITAT Woods, hedges, dry grassland, cultivated areas and dunes.

Water Forget-me-not
Myosotis scorpioides

SIZE AND DESCRIPTION Up to 12cm tall.
A generally hairless creeping perennial.
Leaves are oblong to lanceolate, and
borne on upright stems. Flowers are up
to 8mm across with five joined petals
that are blue with a central yellow eye;
borne in terminal clusters.
FLOWERING TIME May–September.
DISTRIBUTION Widespread across
northern, western and central Europe.
HABITAT Watery habitats on neutral and
basic soils beside rivers and in marshes.

Common Comfrey
Symphytum officinale

Size and description Up to 1.2m tall. An
upright hairy perennial. Basal leaves are
oval, hairy and up to 25cm long; stem leaves are shorter and often
clasp the stems. Flowers are 13–19mm long, tubular, bell-shaped and
variable in colour, but usually creamy white or pinkish-purple; borne
in curved clusters.

Flowering time May–June.

Distribution Widespread in most of Europe in suitable habitats.

Habitat Damp ground; often beside rivers or in fens and marshes.

Oysterplant
Mertensia maritima

SIZE AND DESCRIPTION Up to 60cm long.
A sprawling perennial. Fleshy bluish-
grey leaves are spoon-shaped and
pointed; upper leaves are stalked,
lower ones stalkless. Flowers are
6mm across, pink at first, turning
pale blue, funnel-shaped and
borne in terminal clusters.
FLOWERING TIME June–August.
DISTRIBUTION Locally
distributed on Atlantic coast
of northern Britain and
Northern Ireland, and on
coast of Europe from Jutland
northwards.
HABITAT Sand and shingle
beaches around high-
tide mark.

Wild Marjoram
Origanum vulgare

SIZE AND DESCRIPTION Up to 90cm tall. An aromatic perennial with tall, erect and branching stems. Leaves are opposite, oval and slightly hairy. Flowers are 6–8mm long, maroon in bud and pinkish-purple when flowering, and arranged in dense terminal clusters.

FLOWERING TIME July–September.

DISTRIBUTION Common and widespread throughout most of Europe.

HABITAT Scrub, grassland, hedge banks and downs on calcareous soils.

White Dead-nettle
Lamium album

SIZE AND DESCRIPTION Up to 80cm tall. A hairy and slightly aromatic perennial. When flowers are absent, resembles the Common Nettle (*Urtica dioica*), but it is unrelated to this species and stingless. Leaves are ovate to heart-shaped, nettle-like and appear as opposite pairs on stems. Flowers are 20–25mm long and white, with a hooded upper lip; borne in dense whorls.

FLOWERING TIME March–December.

DISTRIBUTION Widespread in much of Europe but scarce in south.

HABITAT Roadside verges, hedgerows and woodland margins.

Red Dead-nettle
Lamium purpureum

SIZE AND DESCRIPTION Up to 30cm tall. A branched and spreading downy annual that is aromatic when crushed. Leaves are heart-shaped to oval, round-toothed, stalked, paired on stem and sometimes purple-tinged. Flowers are 12–18mm long and purplish-pink, with a hooded upper lip and lower lip toothed at the base; they form clusters at the bases of upper leaves.

FLOWERING TIME March–October.

DISTRIBUTION Found throughout Europe.

HABITAT Cultivated and waste ground.

Common Hemp-nettle
Galeopsis tetrahit

SIZE AND DESCRIPTION
Up to 80cm tall. An upright and branched hairy-stemmed annual. Leaves are nettle-like, ovate, toothed and stalked, paired on stem. Flowers are 15–20mm long and pinkish, striped and spotted with deeper purple; five petals join in a tube with hood-like upper lobes, the lower three bent back.
FLOWERING TIME July–September.
DISTRIBUTION Across Europe.
HABITAT Cultivated ground, wood clearings and waste ground.

Yellow Archangel
Lamiastrum galeobdolon

SIZE AND DESCRIPTION Up to 45cm tall. A creeping and patch-forming hairy perennial with upright flowering stems. Leaves are nettle-like, toothed and oval, and borne in opposite pairs. Flowers are 15–25mm long and bright yellow with a hooded upper lip; borne in whorls around stem.

FLOWERING TIME May–June.

DISTRIBUTION Widespread and locally common across much of Europe in suitable habitats.

HABITAT Woodlands, hedgerows and other shady places. Can form sizeable clumps in suitable locations.

Selfheal
Prunella vulgaris

SIZE AND DESCRIPTION Up to 20cm tall.
A creeping, mat-forming downy
perennial that is popular with insects.
Leaves are oval and sometimes bluntly
toothed; borne in opposite pairs. Flowers are 13–15mm long, two-
lipped and violet-blue; borne in short and dense cylindrical heads.
Flower bracts persist after flowers have dropped.

FLOWERING TIME March–November.

DISTRIBUTION Widespread and often common in much of Europe.

HABITAT Meadows and grassy woodland rides located on calcareous
or neutral soils.

Wild Thyme
Thymus polytrichus

SIZE AND DESCRIPTION Up to 5cm
tall. An aromatic creeping
perennial that often forms
extensive carpets. Leaves are ovate to circular and short-stalked; borne
in opposite pairs along wiry stems. Flowers are 7–12mm long, two-
lipped and pinkish-purple; borne in dense terminal heads.

FLOWERING TIME April–August.

DISTRIBUTION Widespread and locally common in most of central,
western and southern Europe.

HABITAT Free-draining sites like dry grassland and heaths; also found
on coastal sand dunes and cliffs.

Water Mint
Mentha aquatica

SIZE AND DESCRIPTION Up to 50cm tall. A hairy perennial with stiff stems that smells strongly of mint. Leaves are paired on stem, oval and toothed. Flowers are 5–8mm across, have five petals and are pinkish-lilac; they form rounded and dense short-stalked clusters at top of stem. Aromatic flowers attract many insects.

FLOWERING TIME July–October

DISTRIBUTION Found throughout the whole of Europe.

HABITAT Riverbanks, ditches, wetlands and moorland meadows.

Wood Sage
Teucrium scorodonia

SIZE AND DESCRIPTION Up to 50cm tall. A striking hairy perennial with upright branched stems. Sometimes forms sizeable clumps. Leaves are stalked, ovate to heart-shaped, wrinkled and sage-like. Flowers are up to 9mm long, greenish-yellow and paired; borne on upright spikes up to 15cm long.

FLOWERING TIME July–September.

DISTRIBUTION Widespread and locally common in western, southern and central parts of Europe as far north as the Baltic.

HABITAT Woodland rides, heaths and coastal cliffs on acid soils.

Bugle
Ajuga reptans

SIZE AND DESCRIPTION Up to 20cm tall. An upright perennial with hairy stems and leafy creeping runners that root at intervals. Leaves are blunt, sometimes toothed and often stalked, with lower leaves forming a rosette and stem leaves being paired. Flowers are 15mm long and bluish-violet, with five petals forming a tube.

FLOWERING TIME May–July.

DISTRIBUTION Almost everywhere in Europe.

HABITAT Woods and grassy places, usually on damp heavy soils.

Hedge Woundwort
Stachys sylvatica

SIZE AND DESCRIPTION Up to 75cm tall. A hairy perennial with creeping stems, upright flower stalks and an unpleasant smell when bruised. Leaves are oval or heart-shaped, pointed and coarsely toothed, and paired on stem. Flowers are 13–18mm long, reddish-purple and grow in small groups clustered into elongated heads.

FLOWERING TIME July–August.

DISTRIBUTION Almost everywhere in Europe.

HABITAT Woods, hedges and other shady places on damp and nutrient-rich soil.

Ground-ivy
Glechoma hederacea

SIZE AND DESCRIPTION Up to 15cm
tall. A distinctively smelling mat-
forming perennial with softly hairy
creeping stems that root at regular
intervals. Leaves are rounded,
long-stalked and coarsely toothed.
Flowers are 15–20mm long and
bluish-violet, paired or in groups
of four at leaf base.
FLOWERING TIME
March–May.
DISTRIBUTION Found
throughout Europe.
HABITAT Bushy or wooded
grassy places on nutrient-
rich soil.

Wild Clary
Salvia verbenaca

SIZE AND DESCRIPTION Up to 80cm tall. An upright downy perennial. Basal leaves are oval, sometimes pinnately lobed and with jagged-toothed margins; they form a rosette. Stem leaves are smaller and, together with the bracts, are purplish. Flowers are 6–10mm long and variable in colour, but usually blue or violet.

FLOWERING TIME
April–September.

DISTRIBUTION Widespread and local across western and southern Europe.

HABITAT Dry grassy habitats, often coastal, invariably on lime-rich soils.

Black Nightshade
Solanum nigrum

SIZE AND DESCRIPTION Up to 70cm tall. A straggling or upright branching annual that may be hairless or downy. Leaves are oval and pointed, and short-stalked. Flowers are 7–12mm across, with five white lobes and yellow anthers that form a projecting cone. Poisonous berries are rounded and 6–10mm across, green at first but ripening to black.

FLOWERING TIME January–October.

DISTRIBUTION Widespread across most of Europe and often common.

HABITAT Disturbed ground and cultivated land, including gardens.

Bittersweet
Solanum dulcamara

Size and description Up to 1.5m tall. A downy perennial that is woody
at the base and has stems that clamber over other plants. Leaves are
oval and pointed, growing on alternate sides of stem. Flowers are
10–15mm across with five purple petals and bright yellow stamens.
Glossy egg-shaped red berries each contain several rounded seeds;
berries are poisonous.

Flowering time June–September.

Distribution Occurs almost everywhere in Europe.

Habitat Hedges, woods, rocks and shingle, and waste ground.

Common Figwort
Scrophularia nodosa

SIZE AND DESCRIPTION Up to 70cm tall. A hairless perennial that is foetid-smelling when rubbed. Stems square in cross-section and typically not winged. Leaves are short-stalked, oval and pointed, with sharp-toothed margins. Flowers are 7–10mm across, two-lipped and green with maroon-brown lips; borne in open branched spikes. FLOWERING TIME June–September. DISTRIBUTION Widespread in central, western and southern Europe. HABITAT Damp woodland, and shady verges and hedgerows.

Great Mullein
Verbascum thapsus

SIZE AND DESCRIPTION Up to 2m tall. A robust biennial covered in woolly white hairs. Basal leaves are elliptical and up to 50cm long; form rosette in first year. Stem leaves have stalks running down to winged stems. Flowers are 12–35mm across with five yellow lobes; borne on dense and usually unbranched spikes.

FLOWERING TIME June–August.

DISTRIBUTION Widespread in most of Europe except far north and south-east.

HABITAT Dry grassy places, roadside verges and waste ground.

Common Toadflax
Linaria vulgaris

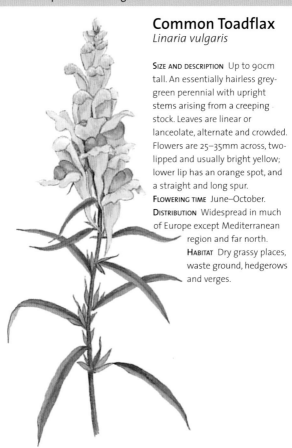

SIZE AND DESCRIPTION Up to 90cm tall. An essentially hairless grey-green perennial with upright stems arising from a creeping stock. Leaves are linear or lanceolate, alternate and crowded. Flowers are 25–35mm across, two-lipped and usually bright yellow; lower lip has an orange spot, and a straight and long spur.

FLOWERING TIME June–October.

DISTRIBUTION Widespread in much of Europe except Mediterranean region and far north.

HABITAT Dry grassy places, waste ground, hedgerows and verges.

Foxglove
Digitalis purpurea

SIZE AND DESCRIPTION Up to 1.8m tall. A tall and elegant biennial or short-lived perennial. Leaves are ovate to lanceolate, long-stalked and softly hairy; in the first year they form a basal rosette, from which the flower spike arises in the second year. Flowers are 40–55mm long, tubular and borne on dense spikes. Very popular with pollinating insects.
FLOWERING TIME June–September.
DISTRIBUTION Widespread across much of western Europe.
HABITAT Woodlands, moors and sea cliffs, usually on acid soils.

Common Cow-wheat
Melampyrum pratense

Size and description Up
to 50cm tall. A variable
hairless or slightly
downy annual that may
be branched and has an
upright habit. A semi-parasite of
grassland plants. Leaves are narrow and
shiny; borne in opposite pairs. Flowers are
10–18mm long, two-lipped, tubular and
bright yellow; flowers arise from leaf axils.
Flowering time May–September.
Distribution Widespread in most of Europe.
Habitat Grassy woodland rides and grassy
heaths mainly on acid soils.

Yellow-rattle
Rhinanthus minor

SIZE AND DESCRIPTION Up to 50cm tall. A branched and hairy or hairless annual that is a characteristic component of many meadows. A semi-parasite of grassland plants. Stems are black-spotted. Leaves are opposite and oblong with rounded teeth. Flowers are 10–20mm long and yellow; borne in leafy terminal spikes. Seeds rattle inside ripe fruits.

FLOWERING TIME May–September.

DISTRIBUTION Common in much of Europe except Mediterranean region.

HABITAT Rough grassy places such as undisturbed meadows.

Marsh Lousewort
Pedicularis palustris

SIZE AND DESCRIPTION Up
to 20cm tall. A spreading
perennial that is branched
from the base. A semi-
parasite of wetland plants.
Leaves are alternate and
feathery, varying from
roughly triangular or
lanceolate to pinnately lobed
and toothed. Flowers are
13–15mm long, pink and two-
lipped, with a straight corolla.
FLOWERING TIME April–July.
DISTRIBUTION Widespread
but local in western and
central Europe.
HABITAT Damp heaths and
moors on nutrient-poor
acid soils.

Common Broomrape
Orobanche minor

SIZE AND DESCRIPTION Up to 40cm tall.
A distinctive upright annual that is parasitic
on the roots of clovers and other herbaceous
plants. Plant totally lacks chlorophyll. Leaves
are reduced to brownish scales on stem.
Flowers are 10–18mm long, two-lipped and
tubular; colour is variable, but usually
pinkish-yellow and purple-veined.
FLOWERING TIME June–September.
DISTRIBUTION Widespread in most of central,
western and southern Europe.
HABITAT Wide range of grassy habitats where
suitable host plants flourish.

Moschatel
Adoxa moschatellina

SIZE AND DESCRIPTION Up to 10cm tall. A diminutive perennial with upright unbranched stems, sometimes forming carpets. Leaves are pale green, fleshy and divided into three leaflets. Flowers are 8–10mm across, yellowish-green and borne on long-stalked heads of five flowers, four facing upwards, the fifth outwards. Fruits are greenish and berry-like.

FLOWERING TIME April–May.

DISTRIBUTION Found in most of Europe.

HABITAT Woods and hedgerows, and among rocks.

Common Valerian
Valeriana officinalis

SIZE AND DESCRIPTION Up to
1.5m tall. An upright perennial with
flowering stems growing from the base. Leaves are spear-shaped, often
toothed and paired. Flowers are 3–5mm long, each with five pinkish-
lilac petals forming a tube; flowers are borne in broad, tightly packed
and branched heads. Fruits are oblong, bearing a feathery parachute.
FLOWERING TIME June–August.
DISTRIBUTION Almost everywhere in Europe.
HABITAT Damp meadows, riverbanks, ditches, damp forests, fens
and scrub.

Greater Bladderwort
Utricularia vulgaris

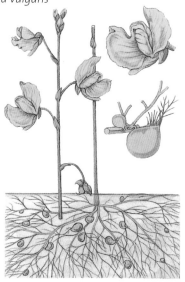

SIZE AND DESCRIPTION Carnivorous aquatic plant with submerged stems up to 1m long, and leafless flowering stems rising up to 20cm above the water. Small bladders along submerged stems trap tiny invertebrates. Leaves are finely divided with bristled teeth. Flowers are 12–18mm long and deep yellow, borne in clusters on stems. Globular capsule splits to release many angular seeds.

FLOWERING TIME July–August.

DISTRIBUTION Occurs throughout Europe.

HABITAT Ponds, lakes and ditches.

Heath Speedwell
Veronica officinalis

SIZE AND DESCRIPTION Up to 10cm tall. Low-growing
perennial with mat-forming habit. Prostrate stems root
at nodes and are hairy all round. Leaves are stalked,
ovate to elliptical, toothed and softly hairy. Flowers are
8mm across, lilac-blue with
dark veins, and comprise four
equal sepal lobes and four
unequal petal lobes.
FLOWERING TIME May–August.
DISTRIBUTION Widespread and
locally common throughout
most of Europe.
HABITAT Grassy woodland rides
and dry heathland areas.

Common Field-speedwell
Veronica persica

Size and description Up to 40cm tall. Prostrate annual with branched stems that angle upwards. Leaves are pale green, oval, coarsely toothed and paired. Flowers are 6–8mm across and mainly sky blue, but with white on lower petal.

Flowering time January–December.

Distribution Widespread and common, although probably not native and introduced from Asia.

Habitat Arable fields, gardens and disturbed ground.

Brooklime
Veronica beccabunga

SIZE AND DESCRIPTION Up to 30cm tall. A hairless perennial with creeping and rooting stems, which become upright. Leaves are oval and fleshy, and borne on short stalks. Flowers are 7–8mm across and blue with a white centre; borne in pairs arising from leaf axils.

FLOWERING TIME May–September.

DISTRIBUTION Widespread in suitable habitats across Europe except far north and drier Mediterranean regions.

HABITAT Restricted to shallow water and damp soil beside rivers and ponds.

Greater Plantain
Plantago major

SIZE AND DESCRIPTION Up to 20cm tall. A perennial often found on lawns, consisting of a tough leafy rosette that sends up flowering spikes. Leaves are broad and oval with a distinct narrow stalk. Flowers are 3mm across with a pale yellow corolla and purple anthers when young, turning yellow with age. Fruits are oblong capsules; tops detach to release seeds.

FLOWERING TIME May–September.

DISTRIBUTION Common and widespread.

HABITAT Lawns, disturbed grassland and arable land.

Ribwort
Plantago lanceolata

SIZE AND DESCRIPTION Up to 15cm tall. Perennial forming a tuft of long slender leaves and long-stalked flowerheads. Leaves are spear-shaped in spreading basal rosettes. Flowers are 4mm across with four brownish petals and four long protruding stamens. Fruits are oblong capsules; tops detach to release seeds.

FLOWERING TIME April–August.

DISTRIBUTION Common and widespread.

HABITAT Disturbed grassland, cultivated ground and tracks.

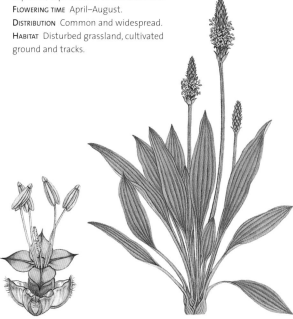

Honeysuckle
Lonicera periclymenum

Size and description Up to 6m long. A vigorous woody deciduous climber that often twines through the branches of trees and shrubs. Leaves are oblong to elliptical, dark green above but greyish-green below; borne in opposite pairs. Flowers are 35–55mm long and creamy white suffused with red or purple; borne on long-stalked terminal heads. Flowers are sweet-scented. Fruits are red berries.

Flowering time June–October.

Distribution Widespread in western, central and southern Europe.

Habitat Woodlands and hedgerows.

Wild Teasel
Dipsacus fullonum

Size and description
Up to 2m tall.
A hairless biennial
whose stout stems are
armed with sharp prickles.
Produces rosettes of elliptical
or oblong spine-covered leaves
in first year. In second year, tall
branching stems with narrower
stem leaves are produced.
Flowers are purple and borne in
spiny conical heads, which
persist in seed.
Flowering time July–August.
Distribution Widespread
throughout western, southern
and central Europe.
Habitat Grassland on damp
disturbed soils.

Field Scabious
Knautia arvensis

SIZE AND DESCRIPTION Up
to 1m tall. A hairy biennial.
Basal leaves are pinnately lobed or
entire and form a basal rosette. Stem
leaves are pinnately divided into up to 16 narrow lobes and a broader
terminal one. Flowers are small, and pink or lilac, and petals of outer
flowers are larger than those of inner ones; borne in flat-topped
heads 15–40mm across.

FLOWERING TIME June–October.

DISTRIBUTION Most of Europe except Mediterranean region.

HABITAT Grassy places, usually only on calcareous soils.

Harebell
Campanula rotundifolia

SIZE AND DESCRIPTION Up to
40cm tall. A delicate hairless
perennial with wiry
stems. Leaves alternate
on stem, and are
circular at base with
rounded teeth and a stalk,
and longer, very narrow and
stalkless on top. Flowers are
15mm-long nodding blue
bells with five petals; borne
on long stalks.
FLOWERING TIME July–September.
DISTRIBUTION Widespread and
common in Europe.
HABITAT Grassy places, open
forests, woodland,
roadside verges, heaths
and dunes.

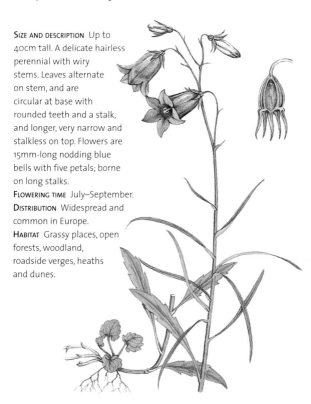

Creeping Bellflower
Campanula rapunculoides

SIZE AND DESCRIPTION Up to 1m tall. A clump-forming upright perennial that may be downy or hairless. Spreads by means of creeping rootstock. Basal leaves are ovate, heart-shaped at the base and long-stalked. Stem leaves are narrower and unstalked. Flowers are 20–30mm long, bell-shaped and bluish-violet; borne in tall spikes.

FLOWERING TIME June–September.

DISTRIBUTION Widespread in central and western Europe; naturalized in Britain.

HABITAT Grassy places including roadside verges and meadows.

Hemp-agrimony
Eupatorium cannabinum

SIZE AND DESCRIPTION Up to 1.75m tall. A distinctive upright downy
perennial. Stems are often reddish. Leaves are divided into 3–5 lobes
and are borne in opposite pairs up the stems. Flowers are dull pink
and small; grouped in dense clusters 2–5mm across, which are borne
in loose terminal inflorescences.

FLOWERING TIME
July–September.

DISTRIBUTION
Widespread in
suitable habitats
across most of
Europe except
far north.

HABITAT Damp ground
such as fens and marshes;
occasionally drier
situations.

Common Ragwort
Senecio jacobaea

SIZE AND DESCRIPTION

Up to 1m tall. An invasive, mostly hairless biennial or perennial. The whole of the plant is highly poisonous; often covered in the orange- and black-striped larvae of the Cinnabar Moth (*Tyria jacobaeae*). Leaves are pinnately divided with a blunt end lobe. Flowerheads are 15–25mm across and bright yellow; borne in flat-topped clusters.

FLOWERING TIME June–November.

DISTRIBUTION Widespread throughout most of Europe.

HABITAT Dry grassy places and verges. Avoided by grazing animals and hence often thrives in pastures.

Scentless Mayweed
Tripleurospermum inodorum

Size and description Up to 75cm
tall. A scentless and hairless
annual or perennial that is much-
branched. Leaves are alternate,
very divided and feathery.
Flowerheads are 20–40mm across and daisy-like, comprising densely
packed small yellow disc florets in centre, surrounded by radiating
longer white ray florets; borne in long-stalked clusters.
Flowering time April–October.
Distribution Widespread across most of Europe in suitable habitats.
Habitat Disturbed ground, cultivated soil, tracks and waste places.
Forms extensive carpets in suitable locations.

Oxeye Daisy
Leucanthemum vulgare

Size and description Up to 60cm tall. A downy or hairless perennial with unbranched upright stems. Leaves are dark green, oval or oblong, toothed or lobed, and arranged spirally on stem. Flowers are 30–50mm across with yellow disc florets and white ray florets.

Flowering time June–August.

Distribution Widespread and common in most of Europe.

Habitat Dry grassy meadows and verges, often on disturbed ground.

Daisy
Bellis perennis

SIZE AND DESCRIPTION Up to 10cm tall. A downy perennial with a leafy rosette sending up leafless flowering stalks. Leaves are spoon-shaped and toothed. Flowers are 15–25mm across with yellow disc florets and white, often pink-tipped ray florets. Fruits are hairy and nut-like.

FLOWERING TIME March–October.

DISTRIBUTION Widespread and common in Europe.

HABITAT Lawns and other areas of short grass.

Tansy
Tanacetum vulgare

SIZE AND DESCRIPTION
Up to 75cm tall.
A robust perennial that
is strongly aromatic and
produces angular upright
stems. Leaves are yellowish-
green with deeply cut lobes,
and are arranged spirally on
stem. Flowers are 7–12mm across,
golden-yellow and button-like,
consisting of disc florets only;
borne in flat-topped clusters.
FLOWERING TIME July–September.
DISTRIBUTION Most of Europe.
HABITAT Hedges, roadside verges
and waste ground.

Feverfew
Tanacetum parthenium

Size and description Up to 50cm tall.
An upright perennial that is downy and
aromatic. Leaves are yellowish-green and
pinnately divided; lower ones are long-stalked, upper ones unstalked.
Flowerheads are daisy-like and 10–20mm across, comprising dense
yellow central disc florets surrounded by radiating white ray florets.
Flowering time July–September.
Distribution Native to south-east Europe. Widespread in much of the
rest of Europe except far north.
Habitat Disturbed ground, roadsides and walls.

Colt's-foot
Tussilago farfara

SIZE AND DESCRIPTION Up to 15cm tall.
A perennial best known for its flower
spikes, which appear in early spring. Leaves
are large and round with shallow lobes
and teeth around the margin. Flowerheads
are 15–35mm across and comprise yellow
central disc florets and paler yellow ray
florets; flowers are solitary and borne on
stems with purple scales.
FLOWERING TIME February–April.
DISTRIBUTION Widespread in most of Europe.
HABITAT Bare and often disturbed ground,
usually on damp clayey soils.

Butterbur
Petasites hybridus

SIZE AND DESCRIPTION Up to 50cm tall.
A perennial herb with a creeping
rootstock; flowers appear before leaves
emerge. Leaves are up to 1m across,
and heart-shaped with blunt teeth
around the margin, green above and
downy grey below. Flowerheads are
dense and pinkish-lilac; borne in stout
spikes of up to 130 heads on stems
with purple scales. Male flowers are
larger than female flowers, and are
borne on separate plants.

FLOWERING TIME March–May.

DISTRIBUTION Widespread in much of
Europe, but least common in south
and absent from far north.

HABITAT Damp ground located in
waterside habitats.

Common Fleabane
Pulicaria dysenterica

SIZE AND DESCRIPTION Up to 60cm tall. A woolly branched perennial that is distinctive and showy only when in flower. Leaves are oblong-lanceolate with wavy margins, green above and greyish below. Flowerheads are 15–30mm across with orange-yellow central disc florets and bright yellow outer ray floret; borne in open clusters. Leaves have insecticidal properties.

FLOWERING TIME July–September.

DISTRIBUTION Widespread throughout western, central and southern Europe.

HABITAT Damp ground in meadows, marshes and ditches.

Greater Burdock
Arctium lappa

SIZE AND DESCRIPTION Up to 1m tall. A downy branched perennial that is robust and upright in habit. Leaves are alternate, heart-shaped at the bases and large. Flowerheads are 30–40mm across, egg-shaped and short-stalked, comprising dense purple disc florets and bracts with hooked spiny tips. In seed, these catch in animal fur to assist dispersal.

FLOWERING TIME July–September.

DISTRIBUTION Widespread throughout most of Europe except far north.

HABITAT Dry grassy places, roadside verges and open woodland.

Meadow Thistle
Cirsium dissectum

SIZE AND DESCRIPTION
Up to 80cm tall. An
upright perennial with
a creeping rootstock
and stems that are
ridged, downy and
unwinged. Leaves are oval,
toothed, and green and hairy
above, and white-cottony
below. Flowerheads are
20–25mm across and comprise reddish-purple florets arising from
a ball of dark bracts; solitary and borne on long spineless stalks.
FLOWERING TIME June–August.
DISTRIBUTION Widespread in western Europe, but local and declining.
HABITAT Restricted to damp peaty soils in bogs and meadows.

Spear Thistle
Cirsium vulgare

SIZE AND DESCRIPTION Up to 3m tall. An upright branched biennial with stems that are cottony, winged and armed with sharp spines between the leaves. Leaves are pinnately lobed and spiny, and often downy beneath. Flowerheads are 20–40mm across and comprise purple florets arising from a ball of spiny bracts. Frequently considered a persistent weed.

FLOWERING TIME July–September.

DISTRIBUTION Widespread across most of Europe.

HABITAT Disturbed ground; often common on wasteland and tracks.

Smooth Sow-thistle
Sonchus oleraceus

Size and description Up to 1m tall. An upright hairless annual or biennial with hollow stems that exude milky sap if broken. The matt grey-green leaves have triangular lobes and spiny margins, and grow from the base or are spirally arranged on the stem. Flowers are yellow and dandelion-like. Fruits are nut-like and have a parachute of hairs.

Flowering time June–August.

Distribution Found throughout Europe.

Habitat Cultivated and waste ground.

Chicory
Cichorium intybus

SIZE AND DESCRIPTION
Up to 1m tall. An
upright branched
perennial with grooved
stems; plant can be hairy or
hairless. Easily overlooked
when flowers are not open.
Lower leaves are stalked and lobed, while
upper ones are narrow and clasping.
Flowerheads are 20–30mm across and sky
blue, and open only on sunny mornings.
FLOWERING TIME July–October.
DISTRIBUTION Widespread in much of Europe in suitable habitats.
HABITAT Dry grassy places; often locally common on roadside verges.

Lesser Hawkbit
Leontodon saxatilis

SIZE AND DESCRIPTION Up to 40cm
tall. An unbranched perennial with
stems that are hairless above and
bristly below. Leaves are pinnately
lobed with wavy-toothed margins.
Flowerheads are 20–25mm across
with yellow florets; flowers are
solitary and borne on unbranched
leafless stalks. Seeds are wind dispersed.
FLOWERING TIME June–October.
DISTRIBUTION Widespread in much of
western, southern and central Europe.
HABITAT Dry grassy habitats, usually on
chalky or sandy soils.

Common Knapweed
Centaurea nigra

Size and description
Up to 1m tall. An
upright perennial that
is roughly hairy and has
grooved stems that
branch towards the top. The
leaves are narrow, and those
near the base of the plant are slightly lobed. Flowerheads are
20–40mm across and comprise purple florets arising from a
ball of brown bracts.

Flowering time June–September.

Distribution Widespread in Europe as far north as southern
Scandinavia; essentially absent from Mediterranean.

Habitat Grassy places including meadows and roadside verges.

Smooth Hawk's-beard
Crepis capillaris

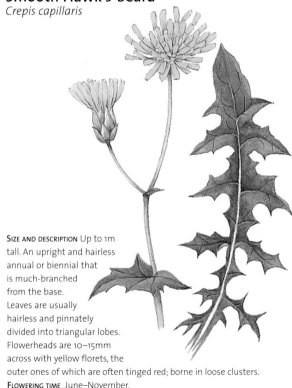

Size and description Up to 1m
tall. An upright and hairless
annual or biennial that
is much-branched
from the base.
Leaves are usually
hairless and pinnately
divided into triangular lobes.
Flowerheads are 10–15mm
across with yellow florets, the
outer ones of which are often tinged red; borne in loose clusters.
Flowering time June–November.
Distribution Widespread in western, central and southern Europe.
Habitat Grassy areas and wasteland including meadows and verges.

Beaked Hawk's-beard
Crepis vesicaria

Size and description Up to 1.2m tall. A bristly branched biennial or perennial with upright grooved stems that exude a milky sap if broken. Leaves are spear-shaped and lobed, growing from base or spirally arranged on stem. Flowers are 15–25mm across with dandelion-like golden-yellow florets, produced in loose clusters. Fruits are nut-like with a hairy white parachute.

Flowering time May–July.

Distribution Western, central and southern Europe.

Habitat Roads, railways, walls and waste ground.

Common Hawkweed
Hieracium vulgatum

SIZE AND DESCRIPTION
Up to 80cm tall. An
upright perennial with
stems that are usually leafy
and yield a milky sap when
broken. Basal leaves are ovate, stalked,
toothed and arranged as a rosette; stem
leaves are unstalked. Flowerheads are
20–30mm across and comprise yellow
florets; borne in groups of up to 20 heads.
FLOWERING TIME July–September.
DISTRIBUTION Widespread in western and
central Europe, but scarce or absent elsewhere.
HABITAT Woods and banks, to heaths and
shady hedgerows.

Goat's-beard
Tragopogon pratensis

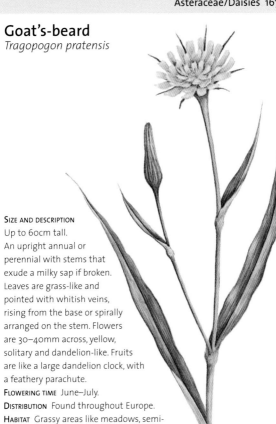

Size and description
Up to 60cm tall.
An upright annual or
perennial with stems that
exude a milky sap if broken.
Leaves are grass-like and
pointed with whitish veins,
rising from the base or spirally
arranged on the stem. Flowers
are 30–40mm across, yellow,
solitary and dandelion-like. Fruits
are like a large dandelion clock, with
a feathery parachute.
Flowering time June–July.
Distribution Found throughout Europe.
Habitat Grassy areas like meadows, semi-
dry grasslands, roadsides and waste ground.

Goldenrod
Solidago virgaurea

SIZE AND DESCRIPTION Up to 75cm tall. An upright perennial with blackish stems. Leaves are oval or elliptical, pointed and toothed, and arranged spirally on stem. Flowers are 5–10mm across, golden, and comprise ray and disc florets; borne on branched spikes. Fruits are nut-like and ribbed with a parachute of hairs.

FLOWERING TIME July–September.

DISTRIBUTION Occurs throughout Europe.

HABITAT Woods, grasslands and rocky banks.

Wild Chamomile
Chamaemelum nobile

SIZE AND DESCRIPTION
Up to 15cm tall. An
aromatic perennial with
prostrate stems, which
sends up ascending
branches with alternate
feathery leaves. Can be
distinguished from closely
related species by its sweet, apple-like
scent and the absence of down on the
undersides of the leaves. Flowers are
18–24mm across with white ray florets
and yellow disk florets, and are borne
singly on long stems.

FLOWERING TIME June–August.

DISTRIBUTION Widespread in Europe.

HABITAT Short grassland, heaths
and waste ground on sandy soils.

Common Cudweed

Filago vulgaris

SIZE AND DESCRIPTION Up to 25cm tall. A woolly greyish-white annual with upright stems. Leaves are spear-shaped and spirally arranged on stem. Flowers are daisy-like, 4–5mm across, growing in compact clusters of 20–35. Fruits are nut-like with a parachute of hairs.

FLOWERING TIME July–August.

DISTRIBUTION Widespread throughout Europe.

HABITAT Grassy places, waste ground and heaths.

Mugwort
Artemisia vulgaris

SIZE AND DESCRIPTION Up to 1.25m tall. An upright aromatic perennial with grooved reddish stems that are branched above. Leaves are deeply lobed, dark green above, silvery and with dense hairs below; lower leaves are stalked, upper ones unstalked. Flowers are 2–3mm across with reddish heads.

FLOWERING TIME July–September.

DISTRIBUTION Widespread throughout Europe.

HABITAT Waste ground, hedges and roadsides.

Yarrow
Achillea millefolium

SIZE AND DESCRIPTION Up to 50cm tall. An upright perennial with leaves that are strongly aromatic and downy. Leaves are feathery and arranged spirally on stem. Flowers are borne in flat-topped clusters of daisy-like white or pink flowerheads, each 4–6mm wide. Fruits are flattened, shiny and nut-like.

FLOWERING TIME June–August.

DISTRIBUTION Found across Europe.

HABITAT Grassy places, meadows, verges, hedgerows and wasteland.

Common Dandelion
Taraxacum officinale

SIZE AND DESCRIPTION Up
to 35cm tall. A familiar
perennial forming a tuft
of leaves and smooth
flowering stems that
exude a milky sap if
broken. Leaves are
oblong or spear-shaped,
and sharply lobed. Flowers
are 20–30mm across and
bright yellow. Fruits, each
with a parachute of hairs,
form dandelion clocks that
are dispersed by the wind.
FLOWERING TIME
March–October.
DISTRIBUTION Common and
widespread in Europe.
HABITAT Grassy places,
gardens and waste ground.

Ramsons
Allium ursinum

Size and description
Up to 50cm tall. An
upright bulbous perennial
that smells strongly of garlic
when bruised. Stem is sharply
angled. Each bulb produces
two bright green leaves that are
up to 25cm long, stalked and
elongated-ovate. Flowers are up
to 20mm across with six white
segments; borne in long-stalked
flat-topped umbels of up to 25
flowers, up to 60mm across.

Flowering time April–June.

Distribution Widespread in most
of Europe except Mediterranean and north-east.

Habitat Woodland, usually on damp or calcareous
soils. Often locally abundant and carpet-forming.

Bluebell
Hyacinthoides non-scripta

SIZE AND DESCRIPTION Up to 50cm
tall. A bulbous perennial. Leaves are
long, narrow and bright green; up to
30cm long with 3–6 arising from base
of plant. Flowers are 14–20mm long,
bluish-purple and bell-shaped,
comprising six segments that are
fused at the base.
FLOWERING TIME April–June.
DISTRIBUTION Widespread in western
Europe; parts of central Europe.
HABITAT Open woodland, often where
coppicing management has taken place;
also hedgerows and sea cliffs. Forms
extensive carpets in suitable locations.

Herb-Paris
Paris quadrifolia

SIZE AND DESCRIPTION Up to 40cm tall. An unusual upright perennial with a creeping rootstock. Upright stems carry a single whorl of unstalked oval to diamond-shaped leaves that are 5–16cm long and conspicuously veined. Flower is solitary, borne on a long stalk, and comprises 4–6 green sepals and four or more narrow petals topped by a purple ovary and yellow stamens.

FLOWERING TIME May–July.

DISTRIBUTION Widespread in most of Europe except Mediterranean.

HABITAT Damp woods on calcareous soils.

Yellow Flag
Iris pseudacorus

SIZE AND DESCRIPTION Up to 1.2m tall. A striking perennial with a large fleshy rhizome. Leaves are sword-shaped, bluish-green and up to 1m long. Stems are slightly flattened. Flowers are up to 100mm across, deep yellow and comprise six segments, the three outer ones being broad and long; borne in groups of 2–3.

FLOWERING TIME June–July.

DISTRIBUTION Widespread across most of Europe.

HABITAT Damp soils. River and pond margins, and water meadows.

Lords-and-ladies
Arum maculatum

SIZE AND DESCRIPTION Up to 25cm tall. An upright and hairless perennial. Leaves are arrowhead-shaped, long-stalked and shiny green, but sometimes purple-spotted; appear in early spring before flowers. Flowers are borne on a spike, male flowers above female ones, but the whole hidden at base of yellow-green spathe and below purple-brown cylindrical spadix. Ripe berries are bright red and appear in autumn.

FLOWERING TIME April–May.

DISTRIBUTION Widespread in western, central and southern Europe.

HABITAT Favours woodland and shady hedgerows, usually on damp soils.

Common Spotted-orchid
Dactylorhiza fuchsii

SIZE AND DESCRIPTION Up to 65cm
tall. An upright perennial. Has
5–12 leaves that are shiny green
and dark-spotted; upper ones
are shorter and narrower
than lower ones. Flowers are
12–18mm across and pinkish-
purple; the lower three-
lobed lip is marked with
dark spots and lines;
borne in tall dense spikes.

FLOWERING TIME
June–August.

DISTRIBUTION Widespread in most of Europe except southern regions.

HABITAT Grassy habitats, usually on chalky or neutral soils. Often
locally abundant.

Common Twayblade
Listera ovata

SIZE AND DESCRIPTION Up to 60cm tall. A perennial orchid with leafy stems. Leaves are broad, rough and oval with several longitudinal veins. Flowers are 14–20mm across, yellowish-green and scented, and 20 or more are borne on each slender flower spike.

FLOWERING TIME June–July.

DISTRIBUTION Almost everywhere in Europe.

HABITAT Woods, hedges and shady grassy places.

Bee Orchid
Ophrys apifera

SIZE AND DESCRIPTION Up to 50cm tall.
A very distinctive perennial. Basal
leaves are ovate-lanceolate and form
a rosette; stem leaves are narrower
ovate. Flowers are 12–25mm across
with three pink segments, two
narrow green ones and a maroon and
furry lower one that is expanded to
form a lip. Yellow markings on lip
add to its resemblance to a bee,
serving to attract the specific
species of solitary bee
(*Andrena hattorfiana*)
that pollinates the
flowers. The flowers
are borne on spikes.
FLOWERING TIME June–July.
DISTRIBUTION Widespread
across much of Europe.
HABITAT Dry grassland,
usually on chalky soils.

Greater Butterfly-orchid
Platanthera chlorantha

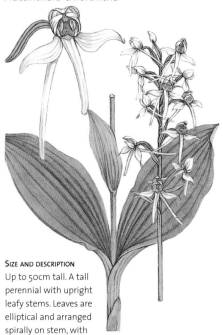

SIZE AND DESCRIPTION
Up to 50cm tall. A tall
perennial with upright
leafy stems. Leaves are
elliptical and arranged
spirally on stem, with
two leaves being up to 150mm long, and smaller leaves above.
Flowers are 18–23mm long, greenish-white, with a distinctive scent.
FLOWERING TIME May–July.
DISTRIBUTION Widespread throughout Europe.
HABITAT Woods and grassy places.

Early-purple Orchid
Orchis mascula

SIZE AND DESCRIPTION Up to 40cm tall. An upright perennial with 3–5 elongated, oval and pointed leaves that are usually spotted . Flowers are 8–12mm long and purplish-crimson, with a three-lobed lower lip and a long nectar spur; they are arranged on a tall loose spike.
FLOWERING TIME April–June.
DISTRIBUTION Common in central and southern Europe; rarer in Scotland and Ireland.
HABITAT Woodland, scrub and grassland on neutral soils.

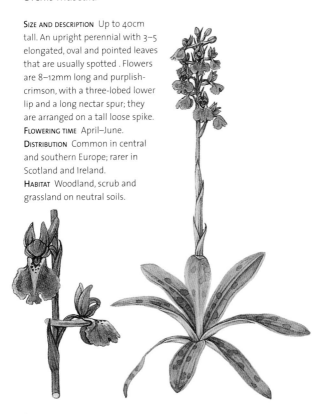

Yellow Water-lily
Nymphaea lutea

SIZE AND DESCRIPTION A striking aquatic perennial with oval leathery floating leaves that are heart-shaped at the bases and up to 40cm across. It also has thinner wavy-edged submerged leaves. Flowers are up to 60mm across with yellow overlapping sepals hiding the petals; borne on slender stalks above the water.

FLOWERING TIME June–September.

DISTRIBUTION Widespread across lowland Europe in suitable habitats.

HABITAT Still or slow-flowing nutrient-rich water. Forms extensive carpets over the water's surface in suitable locations.

White Water-lily
Nymphaea alba

SIZE AND DESCRIPTION A floating plant growing in still or slow-flowing fresh water to a depth of 3m. Leaves are 10–30cm across and rounded. Floating flowers are 150–200mm across, cup-shaped and fragrant. Their 20–25 white or pinkish-white petals open fully only in bright sunshine. Fruits are globular, green and warty, splitting underwater to release many floating seeds.

FLOWERING TIME June–August.

DISTRIBUTION Widespread and locally common in lowland Europe in suitable habitats.

HABITAT Lakes and ponds.

Arrowhead
Sagittaria sagittifolia

Size and description Up to 90cm tall. An upright and hairless aquatic perennial. Aerial leaves are shaped like arrowheads and borne on long upright stalks. Plant also has floating and submerged leaves. Flowers are 15–20mm across with three white petals, each with a purple spot at the base; borne in whorls.

Flowering time July–August.

Distribution Widespread throughout most of Europe except far north and Mediterranean.

Habitat Still or slow-moving waters.

Water-plantain
Alisma plantago-aquatica

SIZE AND DESCRIPTION Up to 1m tall. A waterside perennial forming a leafy tuft with upright flowering stems. Leaves are oval or spear-shaped with a pointed blade. Flowers are pale lilac or white, 8–10mm across and form a head with branches.

FLOWERING TIME June–August.

DISTRIBUTION Found throughout Europe in suitable locations.

HABITAT Edges of ponds, lakes, rivers and canals.

Index

Scientific names